I0065317

A Comprehensive Introduction to Cryobiology

A Comprehensive Introduction to Cryobiology

Edited by
Colby Gunn

Larsen & Keller
www.larsen-keller.com

A Comprehensive Introduction to Cryobiology
Edited by Colby Gunn
ISBN: 978-1-63549-081-7 (Hardback)

Published by Larsen and Keller Education,
5 Penn Plaza,
19th Floor,
New York, NY 10001, USA

Cataloging-in-Publication Data

A comprehensive introduction to cryobiology / edited by Colby Gunn.
p. cm.
Includes bibliographical references and index.
ISBN 978-1-63549-081-7
1. Cryobiology. 2. Cryopreservation of organs, tissues, etc. I. Gunn, Colby.
QH324.9.C7 C66 2017
570.752--dc23

Printed and bound in the United States of America.

For more information regarding Larsen and Keller Education and its products, please visit the publisher's website www.larsen-keller.com

Table of Contents

Preface VII

Chapter 1 **Introduction to Cryobiology** 1

Chapter 2 **Cryopreservation: An Overview** 8
- i. Cryopreservation 8
- ii. Liquid Nitrogen 36
- iii. Glass Transition 41
- iv. Ex Situ Conservation 50
- v. Cryoprotectant 55
- vi. Cryostasis (Clathrate Hydrates) 57
- vii. Neuropreservation 58

Chapter 3 **Cryopreservation in Nature** 60
- i. Antifreeze Protein 60
- ii. Antifreeze 66
- iii. Psychrophile 73
- iv. Insect Winter Ecology 75

Chapter 4 **Cryogenics: An Overview** 82
- i. Cryogenics 82
- ii. Cryogenic Deflashing 87
- iii. Cryogenic Treatment 89
- iv. Cryogenic Seal 91
- v. Cryogenic Fuel 95
- vi. Cryogenic Energy Storage 98
- vii. Cryogenic Storage Dewar 99
- viii. Crystal 100
- ix. Cryotank 109
- x. Absolute zero 110
- xi. Targeted Temperature Management 118

Chapter 5 **Hibernation** 126
- i. Hibernation 126
- ii. Heterothermy 130
- iii. Hibernaculum (zoology) 132

Chapter 6 **Understanding Hypothermia** 133
- i. Hypothermia 133
- ii. Chilblains 143
- iii. Frostbite 145
- iv. Trench Foot 149
- v. Thermoregulation 150

Chapter 7 **Applications of Cryobiology** **163**
i. Cloning 163
ii. Molecular Cloning 175
iii. Organ Transplantation 183
iv. Sperm Bank 204
v. Semen Extender 213
vi. In Vitro Fertilisation 216
vii. Embryo Transfer 240
viii. Cryosurgery 248
ix. Cryoablation 251

Permissions

Index

Preface

The book aims to shed light on some of the unexplored aspects of cryobiology. It discusses thoroughly the various basic and fundamental concepts of this subject. Cryobiology, as a part of biology, deals with the study of changes occurring in living beings present in the cryosphere of the atmosphere, when they are subjected to very low temperature. The systems studied in this field are organs, proteins, tissues, cells, and organisms. This text presents the complex subject of cryobiology in the most comprehensible and easy to understand language. It explores all the important aspects of the area in the present day scenario. This textbook is meant for students who are looking for an elaborate reference text on cryobiology.

A detailed account of the significant topics covered in this book is provided below:

Chapter 1- The branch of biology that concerns itself with the reaction of organisms to low temperatures is known as cryobiology. In the etymological sense of the word, cryobiology means cold. The materials which are studied under the subject of cryobiology are proteins, cells, tissues as well as whole organisms. This chapter is an overview of the subject matter incorporating all the major aspects of cryobiology.

Chapter 2- Cryopreservation is the process of cooling of tissues, cells or organs to save them from damage by cooling them. When the temperature is low enough, it efficiently discontinues any damage that can be instigated by enzymatic or chemical activity. This section explains to the reader the importance of cryopreservation through its various uses such as the preservation of semen and eggs in in vitro treatments, and of living tissues through cryostasis.

Chapter 3- Fishes and insects residing in the Arctic and Antarctic manufacture proteins as protection to minimize damage during the cold weather. This substance that is used to protect them is known as cryoprotectant. Freezing colds are tolerated by organisms through cold hardening procedures.

Chapter 4- Cryogenics is the study of materials and their characteristics at very low temperatures. Refrigeration is a very common example of a cryogenic procedure. Cryogenic treatments enhance the quality of metals and other material. This chapter lists all the main applications of cryogenics and the processes that are involved.

Chapter 5- Some animals hibernate to escape extremely cold seasons, in order to endure the weather and also because the food is scarce. Some of the animals known for their hibernation are deer mice, skunks, bears and hedgehogs. Usually before entering hibernation, animals store enough food in them to survive. The major components of hibernation are discussed in this section.

Chapter 6- Hypothermia is a condition of the muscles and tissues of the living body that has been exposed to very low levels of temperature. Hypothermia is often lethal and persons may lose their external limbs. Other conditions caused by extreme cold temperature are also discussed in this chapter such as chilblains, frostbite and trench foot.

Chapter 7- Cryogenic temperatures have various uses. The application of cryobiological sciences for the preservation of cells and tissues is well known. It also plays an important part in the preservation and recovery of genetic codes of our long lost relatives. Some important applications of cryobiology are cloning, organ transplantation, IVF procedures and cryosurgery.

It gives me an immense pleasure to thank our entire team for their efforts. Finally in the end, I would like to thank my family and colleagues who have been a great source of inspiration and support.

Editor

1

Introduction to Cryobiology

The branch of biology that concerns itself with the reaction of organisms to low temperatures is known as cryobiology. In the etymological sense of the word, cryobiology means cold. The materials which are studied under the subject of cryobiology are proteins, cells, tissues as well as whole organisms. This chapter is an overview of the subject matter incorporating all the major aspects of cryobiology.

Cryobiology

Cryobiology is the branch of biology that studies the effects of low temperatures on living things within Earth's cryosphere or in science. The word cryobiology is derived from the Greek words [kryos], « cold », [bios], « life », and [logos], « word » (hence science). In practice, cryobiology is the study of biological material or systems at temperatures below normal. Materials or systems studied may include proteins, cells, tissues, organs, or whole organisms. Temperatures may range from moderately hypothermic conditions to cryogenic temperatures.

Areas of Study

At least six major areas of cryobiology can be identified: 1) study of cold-adaptation of microorganisms, plants (cold hardiness), and animals, both invertebrates and vertebrates (including hibernation), 2) cryopreservation of cells, tissues, gametes, and embryos of animal and human origin for (medical) purposes of long-term storage by cooling to temperatures below the freezing point of water. This usually requires the addition of substances which protect the cells during freezing and thawing (cryoprotectants), 3) preservation of organs under hypothermic conditions for transplantation, 4) lyophilization (freeze-drying) of pharmaceuticals, 5) cryosurgery, a (minimally) invasive approach for the destruction of unhealthy tissue using cryogenic gases/fluids, and 6) physics of supercooling, ice nucleation/growth and mechanical engineering aspects of heat transfer during cooling and warming, as applied to biological systems. Cryobiology would include cryonics, the low temperature preservation of humans and mammals with the intention of future revival, although this is not part of mainstream cryobiology, depending heavily on speculative technology yet to be invented. Several of these areas of study rely on cryogenics, the branch of physics and engineering that studies the production and use of very low temperatures

Cryopreservation in Nature

Many living organisms are able to tolerate prolonged periods of time at temperatures below the freezing point of water. Most living organisms accumulate cryoprotectants such as antinucleating proteins, polyols, and glucose to protect themselves against frost damage by sharp ice crystals. Most plants, in particular, can safely reach temperatures of −4 °C to −12 °C.

Bacteria

Three species of bacteria, *Carnobacterium pleistocenium*, *Chryseobacterium greenlandensis*. and *Herminiimonas glaciei*, have reportedly been revived after surviving for thousands of years frozen in ice. Certain bacteria, notably *Pseudomonas syringae*, produce specialized proteins that serve as potent ice nucleators, which they use to force ice formation on the surface of various fruits and plants at about −2 °C. The freezing causes injuries in the epithelia and makes the nutrients in the underlying plant tissues available to the bacteria. *Listeria* grows slowly in temperatures as low as -1.5 °C and persists for some time in frozen foods.

Plants

Many plants undergo a process called hardening which allows them to survive temperatures below 0 °C for weeks to months.

Animals

Invertebrates

Nematodes that survive below 0 °C include *Trichostrongylus colubriformis* and *Panagrolaimus davidi*. Cockroach nymphs (*Periplaneta japonica*) survive short periods of freezing at -6 to -8 °C. The red flat bark beetle (*Cucujus clavipes*) can survive after being frozen to -150 °C. The fungus gnat *Exechia nugatoria* can survive after being frozen to -50 °C, by a unique mechanism whereby ice crystals form in the body but not the head. Another freeze-tolerant beetle is *Upis ceramboides*. Insect winter ecology and antifreeze protein. Another invertebrate that is briefly tolerant to temperatures down to -273 °C is the tardigrade.

The larvae of *Haemonchus contortus*, a nematode, can survive 44 weeks frozen at -196 °C.

Vertebrates

For the wood frog (*Rana sylvatica*), in the winter, as much as 45% of its body may freeze and turn to ice. “Ice crystals form beneath the skin and become interspersed among the body’s skeletal muscles. During the freeze, the frog’s breathing, blood flow, and heartbeat cease. Freezing is made possible by specialized proteins and glucose, which prevent intracellular freezing and dehydration.” The wood frog can survive up to 11 days frozen at -4 °C.

Other vertebrates that survive at body temperatures below 0 °C include painted turtles (*Chrysemys picta*), gray tree frogs (*Hyla versicolor*), box turtles (*Terrapene carolina* - 48 hours at -2 °C), spring peeper (*Pseudacris crucifer*), garter snakes (*Thamnophis sirtalis*- 24 hours at -1.5 °C), the chorus frog (*Pseudacris triseriata*), Siberian salamander (*Salamandrella keyserlingii* - 24 hours at -15.3 °C), European common lizard (*Lacerta vivipara*) and Antarctic fish such as *Pagothenia borchgrevinki*. Antifreeze proteins cloned from such fish have been used to confer frost-resistance on transgenic plants.

Hibernating Arctic ground squirrels may have abdominal temperatures as low as −2.9 °C (26.8 °F), maintaining subzero abdominal temperatures for more than three weeks at a time, although the temperatures at the head and neck remain at 0 °C or above.

Applied Cryobiology

Historical Background

Boyle

Cryobiology history can be traced back to antiquity. As early as in 2500 BC, low temperatures were used in Egypt in medicine. The use of cold was recommended by Hippocrates to stop bleeding and swelling. With the emergence of modern science, Robert Boyle studied the effects of low temperatures on animals.

In 1949, bull semen was cryopreserved for the first time by a team of scientists led by Christopher Polge. This led to a much wider use of cryopreservation today, with many organs, tissues and cells routinely stored at low temperatures. Large organs such as hearts are usually stored and transported, for short times only, at cool but not freezing temperatures for transplantation. Cell suspensions (like blood and semen) and thin tissue sections can sometimes be stored almost indefinitely in liquid nitrogen temperature (cryopreservation). Human sperm, eggs, and embryos are routinely stored in fertility research and treatments. Controlled-rate and slow freezing are well established techniques pioneered in the early 1970s which enabled the first human embryo frozen birth (Zoe Leyland) in 1984. Since then, machines that freeze biological samples using programmable steps, or controlled rates, have been used all over the world for human, animal, and cell biology – 'freezing down' a sample to better preserve it for eventual thawing, before it is deep frozen, or cryopreserved, in liquid nitrogen. Such machines are used for freezing oocytes, skin, blood products, embryo, sperm, stem cells, and general tissue preservation in hospitals, veterinary practices, and research labs. The number of live births from 'slow frozen' embryos is some 300,000 to 400,000 or 20% of the estimated 3 million *in vitro* fertilized births. Dr Christopher Chen, Australia, reported the world's first pregnancy using slow-frozen oocytes from a British controlled-rate freezer in 1986.

Cryosurgery (intended and controlled tissue destruction by ice formation) was carried out by James Arnott in 1845 in an operation on a patient with cancer. Cryosurgery is not common.

Low temperature bank, Institute for Problems of Cryobiology and Cryomedicine of the National Academy of Sciences of Ukraine

Preservation Techniques

Cryobiology as an applied science is primarily concerned with low-temperature preservation. Hypothermic storage is typically above 0 °C but below normothermic (32 °C to 37 °C) mammalian temperatures. Storage by cryopreservation, on the other hand, will be in the −80 to −196 °C temperature range. Organs, and tissues are more frequently the objects of hypothermic storage, whereas single cells have been the most common objects cryopreserved.

A rule of thumb in hypothermic storage is that every 10 °C reduction in temperature is accompanied by a 50% decrease in oxygen consumption. Although hibernating animals have adapted mechanisms to avoid metabolic imbalances associated with hypothermia, hypothermic organs, and tissues being maintained for transplantation require special preservation solutions to counter acidosis, depressed sodium pump activity. and increased intracellular calcium. Special organ preservation solutions such as Viaspan (University of Wisconsin solution), HTK, and Celsior have been designed for this purpose. These solutions also contain ingredients to minimize damage by free radicals, prevent edema, compensate for ATP loss, etc.

Cryopreservation of cells is guided by the "two-factor hypothesis" of American cryobiologist Peter Mazur, which states that excessively rapid cooling kills cells by intracellular ice formation and excessively slow cooling kills cells by either electrolyte toxicity or mechanical crushing. During slow cooling, ice forms extracellularly, causing water to osmotically leave cells, thereby dehydrating them. Intracellular ice can be much more damaging than extracellular ice.

For red blood cells, the optimum cooling rate is very rapid (nearly 100 °C per second), whereas for stem cells the optimum cooling rate is very slow (1 °C per minute). Cryoprotectants, such as dimethyl sulfoxide and glycerol, are used to protect cells from freezing. A variety of cell types are protected by 10% dimethyl sulfoxide. Cryobiologists attempt to optimize cryoprotectant concentration (minimizing both ice formation and toxicity) and cooling rate. Cells may be cooled at an optimum rate to a temperature between −30 and −40 °C before being plunged into liquid nitrogen.

Slow cooling methods rely on the fact that cells contain few nucleating agents, but contain naturally occurring vitrifying substances that can prevent ice formation in cells that have been moderately dehydrated. Some cryobiologists are seeking mixtures of cryoprotectants for full vitrification (zero ice formation) in preservation of cells, tissues, and organs. Vitrification methods pose a challenge in the requirement to search for cryoprotectant mixtures that can minimize toxicity.

Cryobiology in Humans

Human gametes and two-, four- and eight-cell embryos can survive cryopreservation at -196 °C for 10 years under well-controlled laboratory conditions.

Cryopreservation in humans with regards to infertility involves preservation of embryos, sperm, or oocytes via freezing. Conception, *in vitro*, is attempted when the sperm is thawed and introduced to the 'fresh' eggs, the frozen eggs are thawed and sperm is placed with the eggs and together they are placed back into the uterus or a frozen embryo is introduced to the uterus. Vitrification has flaws and is not as reliable or proven as freezing fertilized sperm, eggs, or embryos as traditional slow freezing methods because eggs alone are extremely sensitive to temperature. Many researchers are also freezing ovarian tissue in conjunction with the eggs in hopes that the ovarian tissue can be transplanted back into the uterus, stimulating normal ovulation cycles. In 2004, Donnez of Louvain in Belgium reported the first successful ovarian birth from frozen ovarian tissue. In 1997, samples of ovarian cortex were taken from a woman with Hodgkin's lymphoma and cryopreserved in a (Planer, UK) controlled-rate freezer and then stored in liquid nitrogen. Chemotherapy was initiated after the patient had premature ovarian failure. In 2003, after freeze-thawing, orthotopic autotransplantation of ovarian cortical tissue was done by laparoscopy and after five months, reimplantation signs indicated recovery of regular ovulatory cycles. Eleven months after reimplantation, a viable intrauterine pregnancy was confirmed, which resulted in the first such live birth – a girl named Tamara.

Therapeutic hypothermia, e.g. during heart surgery on a "cold" heart (generated by cold perfusion without any ice formation) allows for much longer operations and improves recovery rates for patients.

Scientific Societies

The Society for Cryobiology was founded in 1964 to bring together those from the biological, medical, and physical sciences who have a common interest in the effects of low temperatures on biological systems. As of 2007, the Society for Cryobiology had about 280 members from around the world, and one-half of them are US-based. The purpose of the Society is to promote scientific research in low temperature biology, to improve scientific understanding in this field, and to disseminate and apply this knowledge to the benefit of mankind. The Society requires of all its members the highest ethical and scientific standards in the performance of their professional activities. According to the Society's bylaws, membership may be refused to applicants whose conduct is deemed detrimental to the Society; in 1982, the bylaws were amended explicitly to exclude "any practice or application of freezing deceased persons in the anticipation of their reanimation", over the objections of some members who were cryonicists, such as Jerry Leaf. The Society organizes an annual scientific meeting dedicated to all aspects of low-temperature biology. This international meeting offers opportunities for presentation and discussion of the most up-to-date research in cryobiology, as well as reviewing specific aspects through symposia and workshops. Members are also kept informed of news and forthcoming meetings through the Society newsletter, *News Notes*. The 2011-2012 president of the Society for Cryobiology was John H. Crowe.

The Society for Low Temperature Biology was founded in 1964 and became a registered charity in 2003 with the purpose of promoting research into the effects of low temperatures on all types of organisms and their constituent cells, tissues, and organs. As of 2006, the society had around 130 (mostly British and European) members and holds at least one annual general meeting. The program usually includes both a symposium on a topical subject and a session of free communications on any aspect of low-temperature biology. Recent symposia have included long-term stability, preservation of aquatic organisms, cryopreservation of embryos and gametes, preservation of plants, low-temperature microscopy, vitrification (glass formation of aqueous systems during cooling), freeze drying and tissue banking. Members are informed through the Society Newsletter, which is presently published three times a year.

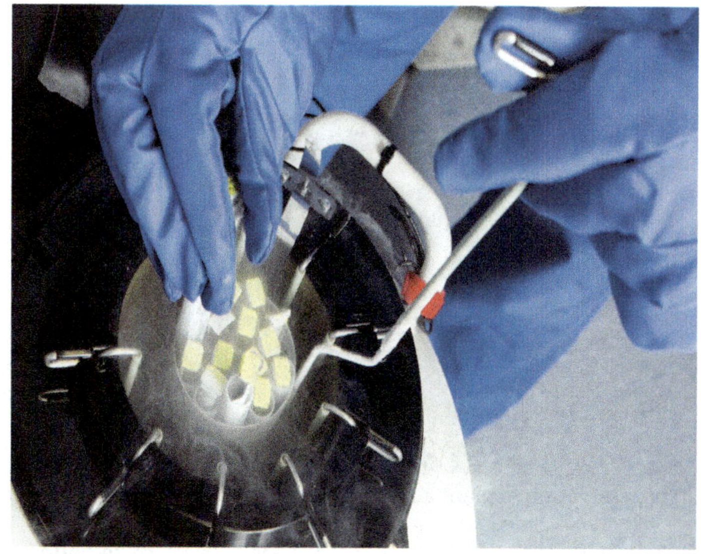

2

Cryopreservation: An Overview

Cryopreservation is the process of cooling of tissues, cells or organs to save them from damage by cooling them. When the temperature is low enough, it efficiently discontinues any damage that can be instigated by enzymatic or chemical activity. This section explains to the reader the importance of cryopreservation through its various uses such as the preservation of semen and eggs in in vitro treatments, and of living tissues through cryostasis.

Cryopreservation

Cryopreservation of plant shoots. Open tank of liquid nitrogen behind.

Cryopreservation or cryoconservation is a process where organelles, cells, tissues, extracellular matrix, organs or any other biological constructs susceptible to damage caused by unregulated chemical kinetics are preserved by cooling to very low temperatures (typically -80 °C using solid carbon dioxide or -196 °C using liquid nitrogen). At low enough temperatures, any enzymatic or chemical activity which might cause damage to the biological material in question is effectively stopped. Cryopreservation methods seek to reach low temperatures without causing additional damage caused by the formation of ice during freezing. Traditional cryopreservation has relied on coating the material to be frozen with a class of molecules termed cryoprotectants. New methods are constantly being investigated due to the inherent toxicity of many cryoprotectants. By default it should be considered that cryopreservation alters or compromises the structure and function of cells unless it is proven otherwise for a particular cell population. Cryoconservation of animal genetic resources is the process in which animal genetic material is collected and stored with the intention of conservation of the breed.

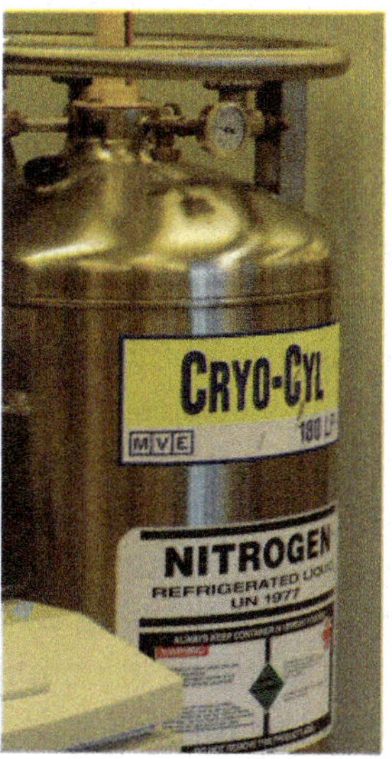

A tank of liquid nitrogen, used to supply a cryogenic freezer (for storing laboratory samples at a temperature of about −150 °C)

Natural Cryopreservation

Water-bears (*Tardigrada*), microscopic multicellular organisms, can survive freezing by replacing most of their internal water with the sugar trehalose, preventing it from crystallization that otherwise damages cell membranes. Mixtures of solutes can achieve similar effects. Some solutes, including salts, have the disadvantage that they may be toxic at intense concentrations. In addition to the water-bear, wood frogs can tolerate the freezing of their blood and other tissues. Urea is accumulated in tissues in preparation for overwintering, and liver glycogen is converted in large quantities to glucose in response to internal ice formation. Both urea and glucose act as "cryoprotectants" to limit the amount of ice that forms and to reduce osmotic shrinkage of cells. Frogs can survive many freeze/thaw events during winter if not more than about 65% of the total body water freezes. Research exploring the phenomenon of "Freezing frogs" has been performed primarily by the Canadian researcher, Dr. Kenneth B. Storey.

Freeze tolerance, in which organisms survive the winter by freezing solid and ceasing life functions, is known in a few vertebrates: five species of frogs (*Rana sylvatica*, *Pseudacris triseriata*, *Hyla crucifer*, *Hyla versicolor*, *Hyla chrysoscelis*), one of salamanders (*Hynobius keyserlingi*), one of snakes (*Thamnophis sirtalis*) and three of turtles (*Chrysemys picta*, *Terrapene carolina*, *Terrapene ornata*). Snapping turtles *Chelydra serpentina* and wall lizards *Podarcis muralis* also survive nominal freezing but it has not been established to be adaptive for overwintering. In the case of *Rana sylvatica* one cryopreservant is ordinary glucose, which increases in concentration by approximately 19 mmol/l when the frogs are cooled slowly.

History

One of the most important early theoreticians of cryopreservation was James Lovelock (born 1919) known for Gaia theory. He suggested that damage to red blood cells during freezing was due to osmotic stress. During the early 1950s, Lovelock had also suggested that increasing salt concentrations in a cell as it dehydrates to lose water to the external ice might cause damage to the cell. Cryopreservation of tissue during recent times began with the freezing of fowl sperm, which during 1957 was cryopreserved by a team of scientists in the UK directed by Christopher Polge. The process was applied to humans during the 1950s with pregnancies obtained after insemination of frozen sperm. However, the rapid immersion of the samples in liquid nitrogen did not, for certain of these samples – such as types of embryos, bone marrow and stem cells – produce the necessary viability to make them usable after thawing. Increased understanding of the mechanism of freezing injury to cells emphasised the importance of controlled or slow cooling to obtain maximum survival on thawing of the living cells. A controlled-rate cooling process, allowing biological samples to equilibrate to optimal physical parameters osmotically in a cryoprotectant (a form of anti-freeze) before cooling in a predetermined, controlled way proved necessary. The ability of cryoprotectants, in the early cases glycerol, to protect cells from freezing injury was discovered accidentally. Freezing injury has two aspects: direct damage from the ice crystals and secondary damage caused by the increase in concentration of solutes as progressively more ice is formed. During 1963, Peter Mazur, at Oak Ridge National Laboratory in the U.S., demonstrated that lethal intracellular freezing could be avoided if cooling was slow enough to permit sufficient water to leave the cell during progressive freezing of the extracellular fluid. That rate differs between cells of differing size and water permeability: a typical cooling rate around 1 °C/minute is appropriate for many mammalian cells after treatment with cryoprotectants such as glycerol or dimethyl sulphoxide, but the rate is not a universal optimum.

Temperature

Storage at very cold temperatures is presumed to provide an indefinite longevity to cells, although the actual effective life is rather difficult to prove. Researchers experimenting with dried seeds found that there was noticeable variability of deterioration when samples were kept at different temperatures – even ultra-cold temperatures. Temperatures less than the glass transition point (Tg) of polyol's water solutions, around −136 °C (137 K; −213 °F), seem to be accepted as the range where biological activity very substantially slows, and −196 °C (77 K; −321 °F), the boiling point of liquid nitrogen, is the preferred temperature for storing important specimens. While refrigerators, freezers and extra-cold freezers are used for many items, generally the ultra-cold of liquid nitrogen is required for successful preservation of the more complex biological structures to virtually stop all biological activity.

Risks

Phenomena which can cause damage to cells during cryopreservation mainly occur during the freezing stage, and include: solution effects, extracellular ice formation, dehydration and intracellular ice formation. Many of these effects can be reduced by cryoprotectants. Once the preserved material has become frozen, it is relatively safe from further damage. However, estimates based on the accumulation of radiation-induced DNA damage during cryonic storage have suggested a maximum storage period of 1000 years.

Solution effects

> As ice crystals grow in freezing water, solutes are excluded, causing them to become concentrated in the remaining liquid water. High concentrations of some solutes can be very damaging.

Extracellular ice formation

> When tissues are cooled slowly, water migrates out of cells and ice forms in the extracellular space. Too much extracellular ice can cause mechanical damage to the cell membrane due to crushing.

Dehydration

> Migration of water, causing extracellular ice formation, can also cause cellular dehydration. The associated stresses on the cell can cause damage directly.

Intracellular ice formation

> While some organisms and tissues can tolerate some extracellular ice, any appreciable intracellular ice is almost always fatal to cells.

Main methods to Prevent Risks

The main techniques to prevent cryopreservation damages are a well established combination of *controlled rate and slow freezing* and a newer flash-freezing process known as *vitrification*.

Slow Programmable Freezing

Controlled-rate and slow freezing, also known as *slow programmable freezing (SPF)*, is a set of well established techniques developed during the early 1970s which enabled the first human embryo frozen birth Zoe Leyland during 1984. Since then, machines that freeze biological samples using programmable sequences, or controlled rates, have been used all over the world for human, animal and cell biology – “freezing down” a sample to better preserve it for eventual thawing, before it is frozen, or cryopreserved, in liquid nitrogen. Such machines are used for freezing oocytes, skin, blood products, embryo, sperm, stem cells and general tissue preservation in hospitals, veterinary practices and research laboratories around the world. As an example, the number of live births from frozen embryos 'slow frozen' is estimated at some 300,000 to 400,000 or 20% of the estimated 3 million in vitro fertilisation (IVF) births.

Lethal intracellular freezing can be avoided if cooling is slow enough to permit sufficient water to leave the cell during progressive freezing of the extracellular fluid. To minimize the growth of extracellular ice crystal growth and recrystallization, biomaterials such as alginates, poly vinyl alcohol or chitosan can be used to impede ice crystal growth along with traditional small molecule cryoprotectants. That rate differs between cells of differing size and water permeability: a typical cooling rate of about 1 °C/ minute is appropriate for many mammalian cells after treatment with cryoprotectants such as glycerol or dimethyl sulphoxide, but the rate is not a universal optimum. The 1 °C / minute rate can be achieved by using devices such as a rate-controlled freezer or a benchtop portable freezing container.

Several independent studies have provided evidence that frozen embryos stored using slow-freezing techniques may in some ways be 'better' than fresh in IVF. The studies indicate that using frozen embryos and eggs rather than fresh embryos and eggs reduced the risk of stillbirth and premature delivery though the exact reasons are still being explored.

Vitrification

Researchers Greg Fahy and William F. Rall helped introduce vitrification to reproductive cryopreservation in the mid-1980s. As of 2000, researchers claim vitrification provides the benefits of cryopreservation without damage due to ice crystal formation. The situation became more complex with the development of tissue engineering as both cells and biomaterials need to remain ice-free to preserve high cell viability and functions, integrity of constructs and structure of biomaterials. Vitrification of tissue engineered constructs was first reported by Lilia Kuleshova, who also was the first scientist to achieve vitrification of woman's eggs (oocytes), which resulted in live birth in 1999. For clinical cryopreservation, vitrification usually requires the addition of cryoprotectants prior to cooling. The cryoprotectants act like antifreeze: they decrease the freezing temperature. They also increase the viscosity. Instead of crystallizing, the syrupy solution becomes an amorphous ice—it *vitrifies*. Rather than a phase change from liquid to solid by crystallization, the amorphous state is like a "solid liquid", and the transformation is over a small temperature range described as the "glass transition" temperature.

Vitrification of water is promoted by rapid cooling, and can be achieved without cryoprotectants by an extremely rapid decrease of temperature (megakelvins per second). The rate that is required to attain glassy state in pure water was considered to be impossible until 2005.

Two conditions usually required to allow vitrification are an increase of the viscosity and a decrease of the freezing temperature. Many solutes do both, but larger molecules generally have a larger effect, particularly on viscosity. Rapid cooling also promotes vitrification.

For established methods of cryopreservation, the solute must penetrate the cell membrane in order to achieve increased viscosity and decrease freezing temperature inside the cell. Sugars do not readily permeate through the membrane. Those solutes that do, such as dimethyl sulfoxide, a common cryoprotectant, are often toxic in intense concentration. One of the difficult compromises of vitrifying cryopreservation concerns limiting the damage produced by the cryoprotectant itself due to cryoprotectant toxicity. Mixtures of cryoprotectants and the use of ice blockers have enabled the Twenty-First Century Medicine company to vitrify a rabbit kidney to -135 °C with their proprietary vitrification mixture. Upon rewarming, the kidney was transplanted successfully into a rabbit, with complete functionality and viability, able to sustain the rabbit indefinitely as the sole functioning kidney.

Freezable Tissues

Generally, cryopreservation is easier for thin samples and small clumps of individual cells, because these can be cooled more quickly and so require lesser doses of toxic cryoprotectants. Therefore, cryopreservation of human livers and hearts for storage and transplant is still impractical.

Nevertheless, suitable combinations of cryoprotectants and regimes of cooling and rinsing during warming often allow the successful cryopreservation of biological materials, particularly cell suspensions or thin tissue samples. Examples include:

- Semen in semen cryopreservation
- Blood
    - Special cells for transfusion
    - Stem cells. It is optimal in high concentration of synthetic serum, stepwise equilibration and slow cooling.
    - Umbilical cord blood *Further information: Cord blood bank#Cryopreservation*
- Tissue samples like tumors and histological cross sections
- Eggs (oocytes) in oocyte cryopreservation
- Embryos at cleavage stage (that are 2, 4 or 8 cells) or at blastocyst stage, in embryo cryopreservation
- Ovarian tissue in ovarian tissue cryopreservation
- Plant seeds or shoots may be cryopreserved for conservation purposes.

Additionally, efforts are underway to preserve humans cryogenically, known as cryonics. For such efforts either the brain within the head or the entire body may experience the above process. Cryonics is in a different category from the aforementioned examples, however: while countless cryopreserved cells, vaccines, tissue and other biologial samples have been thawed and used successfully, this has not yet been the case at all for cryopreserved brains or bodies. At issue are the criteria for defining “success”. Proponents of cryonics claim that cryopreservation using present technology, particularly vitrification of the brain, may be sufficient to preserve people in an “information theoretic” sense so that they could be revived and made whole by hypothetical vastly advanced future technology.

Embryos

Cryopreservation for embryos are used for *embryo storage*, e.g. when in vitro fertilization has resulted in more embryos than is currently needed.

Pregnancies have been reported from embryos stored for 16 years. Many studies have evaluated the children born from frozen embryos, or “frosties”. The result has been uniformly positive with no increase of birth defects or development abnormalities. A study of more than 11,000 cryopreserved human embryos had no significant effect of storage time on post-thaw survival for in vitro fertilisation (IVF) or oocyte donation cycles, or for embryos frozen at the pronuclear or cleavage stages. Additionally, the duration of storage did not have any significant effect on clinical pregnancy, miscarriage, implantation, or live birth rate, whether from IVF or oocyte donation cycles. Rather, oocyte age, survival proportion, and number of transferred embryos are predictors of pregnancy outcome.

Ovarian Tissue

Cryopreservation of ovarian tissue is of interest to women who want to preserve their reproductive function beyond the natural limit, or whose reproductive potential is threatened by cancer therapy, for example in hematologic malignancies or breast cancer. The procedure is to take a part of the ovary and perform slow freezing before storing it in liquid nitrogen whilst therapy is undertaken. Tissue can then be thawed and implanted near the fallopian, either orthotopic (on the natural location) or heterotopic (on the abdominal wall), where it starts to produce new eggs, allowing normal conception to occur. The ovarian tissue may also be transplanted into mice that are immunocompromised (SCID mice) to avoid graft rejection, and tissue can be harvested later when mature follicles have developed.

Oocytes

Human *Oocyte cryopreservation* is a new technology in which a woman’s eggs (oocytes) are extracted, frozen and stored. Later, when she is ready to become pregnant, the eggs can be thawed, fertilized, and transferred to the uterus as embryos. Since 1999, when the birth of the first baby from an embryo derived from vitrified-warmed woman's eggs was reported by Kuleshova and co-workers in the journal of Human Reproduction, this concept has been recognized and widespread. This breakthrough in achieving vitrification of woman's oocytes made an important advance in our knowledge and practice of the IVF process, as clinical pregnancy rate is four times higher after oocyte vitrification than after slow freezing. Oocyte vitrification is vital for preservation fertility in young oncology patients and for individuals undergoing IVF who object, either for religious or ethical reasons, to the practice of freezing embryos.

Semen

Semen can be used successfully almost indefinitely after cryopreservation. The longest reported successful storage is 22 years. It can be used for sperm donation where the recipient wants the treatment in a different time or place, or as a means of preserving fertility for men undergoing vasectomy or treatments that may compromise their fertility, such as chemotherapy, radiation therapy or surgery.

Testicular Tissue

Cryopreservation of immature testicular tissue is a developing method to avail reproduction to young boys who need to have gonadotoxic therapy. Animal data are promising, since healthy offsprings have been obtained after transplantation of frozen testicular cell suspensions or tissue pieces. However, none of the fertility restoration options from frozen tissue, i.e. cell suspension transplantation, tissue grafting and in vitro maturation (IVM) has proved efficient and safe in humans as yet.

Moss

Four different ecotypes of *Physcomitrella patens* stored at the IMSC.

Cryopreservation of whole moss plants, especially Physcomitrella patens, has been developed by Ralf Reski and coworkers and is performed at the International Moss Stock Center. This biobank collects, preserves, and distributes moss mutants and moss ecotypes.

Mesenchymal Stromal Cells (MSCs)

Scientists have reported that MSCs when transfused immediately within few hours post thawing may show reduced function or show decreased efficacy in treating diseases as compared to those MSCs which are in log phase of cell growth(fresh), so cryopreserved MSCs should be brought back into log phase of cell growth in *in vitro* culture before these are administered for clinical trials or experimental therapies, re-culturing of MSCs will help in recovering from the shock the cells get during freezing and thawing. Various clinical trials on MSCs have failed which used cryopreserved product immediately post thaw as compared to those clinical trials which used fresh MSCs [Francois M et al., Cytotherapy.2012;14:147–152].

Preservation of Microbiology Cultures

Bacteria and fungi can be kept short-term (months to about a year, depending) refrigerated, however, cell division and metabolism is not completely arrested and thus is not an optimal option for long-term storage (years) or to preserve cultures genetically or phenotypically, as cell divisions can lead to mutations or sub-culturing can cause phenotypic changes. A preferred option, species-dependent, is cryopreservation.

Fungi

Fungi, notably zygomycetes, ascomycetes and higher basidiomycetes, regardless of sporulation, are able to be stored in liquid nitrogen or deep-frozen. Crypreservation is a hallmark method for fungi that do not sporulate (otherwise other preservation methods for spores can be used at lower costs and ease), sporulate but have delicate spores (large or freeze-dry sensitive), are pathogenic (dangerous to keep metabolically active fungus) or are to be used for genetic stocks (ideally to have identical composition as the original deposit). As with many other organisms, cryoprotectants like DMSO or glycerol (e.g. filamentous fungi 10% glycerol or yeast 20% glycerol) are used. Differences between choosing cryoprotectants are species (or class) dependent, but generally for fungi penetrating cryoprotectants like DMSO, glycerol or polyethylene glycol are most effective (other non-penetrating ones include sugars mannitol, sorbitol, dextran, etc.). Freeze-thaw repetition is not recommended as it can decrease viability. Back-up deep-freezers or liquid nitrogen storage sites are recommended. Multiple protocols for freezing are summarized below (each uses screw-cap polypropylene cryotubes):

A) Non-sporulating fungi or embedded mycelia: 10% glycerol is added to the tube and agar plugs of fresh culture are added and immediately frozen in liquid-nitrogen vapor (−170 °C). Cultures are thawed at 37 °C and plated.

B) Spores or mycelia from agar plate: 10% glycerol or 5% DMSO spore or mycelia suspension are made and frozen.

C) Liquid mycelia: Mycelia are macerated (not for use with human pathogenic fungi) and mixed to make a final concentration of 10% glycerol or 5% DSMO.

For protocol B and C, stocks are gradually cooled until frozen. Similar thawing and plating as in A.

Bacteria

Many common culturable laboratory strains are deep-frozen to preserve genetically and phenotypically stable, long-term stocks. Sub-culturing and prolonged refrigerated samples may lead to loss of plasmid(s) or mutations. Common final glycerol percentages are 15, 20 and 25. From a fresh culture plate, one single colony of interest is chosen and liquid culture is made. From the liquid culture, the medium is directly mixed with equal amount of glycerol; the colony should be checked for any defects like mutations. All antibiotics should be washed from the culture before long-term storage. Methods vary, but mixing can be done gently by inversion or rapidly by vortex and cooling can vary by either placing the cryotube directly at −80 °C, shock-freezing in liquid nitrogen or gradually cooling and then storing at −80 °C or cooler (liquid nitrogen or liquid nitrogen vapor). Recovery of bacteria can also vary, namely if beads are stored within the tube then the few beads can be used to plate or the frozen stock can be scraped with a loop and then plated, however, since only little stock is needed the entire tube should never be completely thawed and repeated freeze-thaw should be avoided. 100% recovery is not feasible regardless of methodology.

Uses of Cryopreservation

Semen Cryopreservation

Semen cryopreservation (commonly called sperm banking) is a procedure to preserve sperm cells. Semen can be used successfully indefinitely after cryopreservation. For human sperm, the longest reported successful storage is 22 years. It can be used for sperm donation where the recipient wants the treatment in a different time or place, or as a means of preserving fertility for men undergoing vasectomy or treatments that may compromise their fertility, such as chemotherapy, radiation therapy or surgery.

Freezing

The most common cryoprotectant used for semen is glycerol (10% in culture medium). Often sucrose or other di-, trisaccharides are added to glycerol solution. Cryoprotectant media may be supplemented with either egg yolk or soy lecithin, with the two having no statistically significant differences compared to each other regarding motility, morphology, ability to bind to hyaluronate in vitro, or DNA integrity after thawing.

Semen is frozen using either a controlled-rate, slow-cooling method (slow programmable freezing or SPF) or a newer flash-freezing process known as vitrification. Vitrification gives superior post-thaw motility and cryosurvival than *slow programmable freezing*.

Thawing

Thawing at 40 °C seems to result in optimal sperm motility. On the other hand, the exact thawing temperature seems to have only minor effect on sperm viability, acrosomal status, ATP content, and DNA.

Refreezing

In terms of the level of sperm DNA fragmentation, up to three cycles of freezing and thawing can be performed without causing a level of risk significantly higher than following a single cycle of freezing and thawing. This is provided that samples are refrozen in their original cryoprotectant and are not going through sperm washing or other alteration in between, and provided that they are separated by density gradient centrifugation or swim-up before use in assisted reproduction technology.

Effect on Quality

Some evidence suggests an increase in single-strand breaks, condensation and fragmentation of DNA in sperm after cryopreservation. This can potentially increase the risk of mutations in offspring DNA. Antioxidants and the use of well-controlled cooling regimes could potentially improve outcomes.

In long-term follow-up studies, no evidence has been found either of an increase in birth defects or chromosomal abnormalities in people conceived from cryopreserved sperm compared with the general population.

Oocyte Cryopreservation

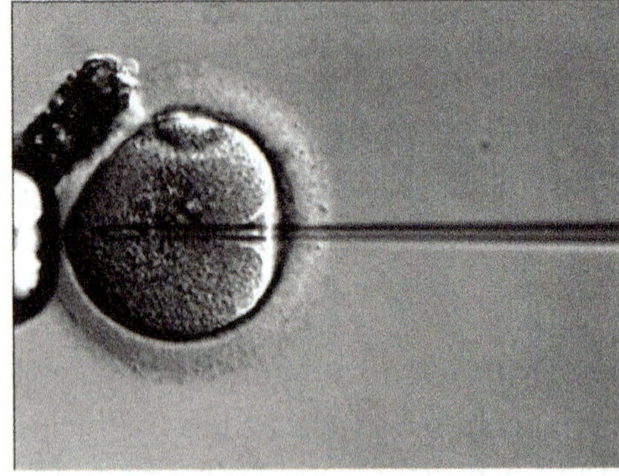

ICSI sperm injection into oocyte

Human oocyte cryopreservation (egg freezing) is a process in which a woman's eggs (oocytes) are extracted, frozen and stored. Later, when she is ready to become pregnant, the eggs can be thawed, fertilized, and transferred to the uterus as embryos.

Indications

Oocyte cryopreservation is aimed at three particular groups of women: those diagnosed with cancer who have not yet begun chemotherapy or radiotherapy; those undergoing treatment with assisted reproductive technologies who do not consider embryo freezing an option; and those who would like to preserve their future ability to have children, either because they do not yet have a partner, or for other personal or medical reasons.

Over 50,000 reproductive-age women are diagnosed with cancer each year in the United States. Chemotherapy and radiotherapy are toxic for oocytes, leaving few, if any, viable eggs. Egg freezing offers women with cancer the chance to preserve their eggs so that they can have children in the future.

Oocyte cryopreservation is an option for individuals undergoing IVF who object, either for religious or ethical reasons, to the practice of freezing embryos. Having the option to fertilize only as many eggs as will be utilized in the IVF process, and then freeze any remaining unfertilized eggs can be a solution. In this way, there are no excess embryos created, and there need be no disposition of unused frozen embryos, a practice which can create complex choices for certain individuals.

There is a thriving industry marketing freezing eggs at an early age, and it may ensure a chance for a future pregnancy.

Additionally, women with a family history of early menopause have an interest in fertility preservation. With egg freezing, they will have a frozen store of eggs, in the likelihood that their eggs are depleted at an early age.

Method

The egg retrieval process for oocyte cryopreservation is the same as that for in vitro fertilization. This includes one to several weeks of hormone injections that stimulate ovaries to ripen multiple eggs. When the eggs are mature, final maturation induction is performed, preferably by using a GnRH agonist rather than human chorionic gonadotrophin (hCG), since it decreases the risk of ovarian hyperstimulation syndrome with no evidence of a difference in live birth rate (in contrast to fresh cycles where usage of GnRH agonist has a lower live birth rate). The eggs are subsequently removed from the body by transvaginal oocyte retrieval. The procedure is usually conducted under sedation. The eggs are immediately frozen.

The egg is the largest cell in the human body and contains a high amount of water. When the egg is frozen, the ice crystals that form can destroy the integrity of the cell. To prevent this, the egg must be dehydrated prior to freezing. This is done using cryoprotectants which replace most of the water within the cell and inhibit the formation of ice crystals.

Eggs (oocytes) are frozen using either a controlled-rate, slow-cooling method or a newer flash-freezing process known as vitrification. Vitrification is much faster but requires higher concentrations of cryoprotectants to be added. The result of vitrification is a solid glass-like cell, free of ice crystals. Indeed, freezing is a phase transition. Vitrification, as opposed to freezing, is a physical transition. Realizing this fundamental difference, vitrification concept has been developed and successfully applied in IVF treatment with the first life birth following vitrification of oocytes achieved in 1999. Vitrification eliminates ice formation inside and outside of oocytes on cooling, during cryostorage and on warming. Vitrification is associated with higher survival rates and better development compared to slow-cooling when applied to oocytes in metaphase II (MII). Vitrification has also become the method of choice for pronuclear oocytes, although prospective randomized controlled trials are still lacking.

During the freezing process, the zona pellucida, or shell of the egg can be modified preventing fertilization. Thus, currently, when eggs are thawed, a special fertilization procedure is performed by an embryologist whereby sperm is injected directly into the egg with a needle rather than allowing sperm to penetrate naturally by placing it around the egg in a dish. This injection technique is called ICSI (Intracytoplasmic Sperm Injection) and is also used in IVF.

Success Rates

The percentage of transferred cycles is lower in frozen cycles compared with fresh cycles (approx. 30% and 50%). Such outcomes are considered comparable.

In a 2013 meta-analysis of more than 2,200 cycles using frozen eggs, scientists found the probability of having a live birth after three cycles was 31.5 percent for women who froze their eggs at age 25, 25.9 percent at age 30, 19.3 percent at age 35, and 14.8 percent at age 40.

Two recent studies showed that the rate of birth defects and chromosomal defects when using cryopreserved oocytes is consistent with that of natural conception.

Recent modifications in protocol regarding cryoprotectant composition, temperature and storage methods have had a large impact on the technology, and while it is still considered an experimental procedure, it is quickly becoming an option for women. Slow freezing traditionally has been the most commonly used method to cryopreserve oocytes, and is the method that has resulted in the most babies born from frozen oocytes worldwide. Ultra-rapid freezing or vitrification represents a potential alternative freezing method.

In the fall of 2009, The American Society for Reproductive Medicine (ASRM) issued an opinion on oocyte cryopreservation concluding that the science holds "great promise for applications in oocyte donation and fertility preservation" because recent laboratory modifications have resulted in improved oocyte survival, fertilization, and pregnancy rates from frozen-thawed oocytes in IVF. The ASRM noted that from the limited research performed to date, there does not appear to be an increase in chromosomal abnormalities, birth defects, or developmental deficits in the children born from cryopreserved oocytes. The ASRM recommended that, pending further research, oocyte cryopreservation should be introduced into clinical practice on an investigational basis and under the guidance of an Institutional Review Board (IRB). As with any new technology, safety and efficacy must be evaluated and demonstrated through continued research.

In October 2012, the ASRM lifted the experimental label from the technology for women with a medical need, citing success rates in live births, among other findings. However, they also warned against using it only to delay child-bearing.

In 2014, a Cochrane systematic review about this topic was published. It compared vitrification (the newest technology) versus slow freezing (the oldest one). Key results of that review showed that the clinical pregnancy rate was almost 4 times higher in the oocyte vitrification group than in the slow freezing group, with moderate quality of evidence.

Cost

The cost of the egg freezing procedure (without embryo transfer) in U.S.U.K and other European countries varies in between $5,000 and $12,000. This does not include the fertility medications which can cost between $4,000 and $5,000. Egg storage can vary from $100 to more than $1,000.

In other countries popular for medical tourism or regional tourism (within the country) , costs may vary. Egg Freezing Costs Worldwide

As an example, in India, which is an upcoming medical tourism country for Infertility & IVF Treatment, egg freezing is catching up, but, there is only one site that publicly has provided costing and it ranges from approximately 700$ (Rs 49,100 INR) for egg freezing including Vitrification cost but excluding fertility medications , as well as approximately 1420$ (Rs 99,100/- INR) for egg freezing with fertility medications. The website claims to have included fertility medications in the cost of Egg Freezing cycle. Cost of Storage is sited as 180$ per year. Egg Freezing Costs in Indian Oocyte Bank

In other countries ,well established in medical tourism and IVF , Czech Republic, Ukraine , Cyprus etc, are also offering Egg Freezing at competitive prices It is sited as a lower cost alternative to typical US options for egg freezing. Spain and the Czech Republic are popular destinations for this treatment.

The cost for egg freezing abroad may start as low as $2500 per cycle (including medications), but you also need to factor in travel costs for 7 – 10 days in Europe for the stimulation and egg collection per cycle. Frozen oocyte storage costs tend to be slightly less expensive as well, with typical costs as low as $200–300 per year. Cost of Egg Freezing at popular Medical Tourism Centres

History

Cryopreservation itself has always played a central role in assisted reproductive technology. With the first cryopreservation of sperm in 1953 and of embryos thirty years later, these techniques have become routine. Dr Christopher Chen of Singapore reported the world's first pregnancy in 1986 using previously frozen oocytes. This report stood alone for several years followed by studies reporting success rates using frozen eggs to be much lower than those of traditional in vitro fertilization (IVF) techniques using fresh oocytes. Providing the lead to a new direction in cryobiology, Dr. Lilia Kuleshova was the first scientist to achieve vitrification of human oocytes that resulted in a live birth in 1999. Then recently, two articles published in the journal, *Fertility and Sterility*, reported pregnancy rates using frozen oocytes that were comparable to those of cryopreserved embryos and even fresh embryos. These newer reports affirm that oocyte cryopreservation technology is advancing.

Almost 42,000 'slow frozen' (as opposed to 'vitrified') human embryo transfers were performed during 2001 in Europe (Andersen et al. 2005). In addition, it is estimated that between 300,000 and 500,000 successful human births have resulted worldwide from the transfer of previously 'slow frozen' embryos performed from the mid-1970s to 2006.

Social Egg Freezing

Social egg freezing is a term used to describe the use of egg-freezing as an attempt to delay child-bearing in a non-medical context. There has been a proliferation in the marketing of this kind of egg freezing since October 2012 when the ASRM lifted the experimental label from the technology, despite their explicit warning against these uses given the risks assumed by the woman and the child.

There was another spike in interest in 2014 when it was revealed that Facebook and Apple were adding egg freezing as a benefit for their female employees. This announcement was controversial, with some women finding it useful or empowering, and others finding it alienating or misguided.

A string of "egg freezing parties" by third-party companies have also helped popularize the concept among young women.

Risks

The risks involved with egg freezing for the women undergoing the procedure are the same as the risks from egg extraction for the purposes of IVF. These include: bleeding from the oocyte recovery procedure, reaction to the hormones used to induce hyperovulation (producing more than one egg), including ovarian hyperstimulation syndrome (OHSS) and, rarely, liver failure. The long-term effects of egg extraction on women's bodies have not been well studied.

There may also be some risks to any resulting child. These will include any of the risks associated with IVF, as well as potential unknown risks caused by long-term freezing, which are currently unknown. Expanded use of Intracytoplasmic sperm injection (ICSI) to inject a single sperm into a thawed egg could additionally be a cause for concern as this method has been associated with a higher rate of birth defects.

Embryo Cryopreservation

Cryopreservation of embryos is the process of preserving an embryo at sub-zero temperatures, generally at an embryogenesis stage corresponding to pre-implantation, that is, from fertilisation to the blastocyst stage.

Indications

Embryo cryopreservation is useful for leftover embryos after a cycle of in vitro fertilisation, as patients who fail to conceive may become pregnant using such embryos without having to go through a full IVF cycle. Or, if pregnancy occurred, they could return later for another pregnancy. Spare oocytes or embryos resulting from fertility treatments may be used for oocyte donation or embryo donation to another woman or couple, and embryos may be created, frozen and stored specifically for transfer and donation by using donor eggs and sperm.

Method

Embryo cryopreservation is generally performed as a component of in vitro fertilization (which generally also includes ovarian hyperstimulation, egg retrieval and embryo transfer). The ovarian hyperstimulation is preferably done by using a GnRH agonist rather than human chorionic gonadotrophin (hCG) for final oocyte maturation, since it decreases the risk of ovarian hyperstimulation syndrome with no evidence of a difference in live birth rate (in contrast to fresh cycles where usage of GnRH agonist has a lower live birth rate).

The main techniques used for embryo cryopreservation are vitrification versus slow programmable freezing (SPF). Studies indicate that vitrification is superior or equal to SPF in terms of survival and implantation rates. Vitrification appears to result in decreased risk of DNA damage than slow freezing.

Direct Frozen Embryo Transfer: Embryos can be frozen by slow programmable freezing (SPF)method in ethylene glycol freeze media and transfer directly to recipients immedieatly after water thawing without laboratory thawing process.The world's first crossbred bovine embryo transfer calf under tropical conditions was produced by such technique on 23 June 1996 by Dr. Binoy S Vettical of Kerala Livestock Development Board, Mattupatti

Prevalence

World usage data is hard to come by but it was reported in a study of 23 countries that almost 42,000 frozen human embryo transfers were performed during 2001 in Europe.

Pregnancy Outcome and Determinants

In current state of the art, early embryos having undergone cryopreservation implant at the same rate as equivalent fresh counterparts. The outcome from using cryopreserved embryos has uniformly been positive with no increase in birth defects or development abnormalities, also between fresh versus frozen eggs used for intracytoplasmic sperm injection (ICSI). In fact, pregnancy rates are increased following frozen embryo transfer, and perinatal outcomes are less affected, compared to embryo transfer in the same cycle as ovarian hyperstimulation was performed. The endometrium is believed to not be optimally prepared for implantation following ovarian hyperstimulation, and therefore frozen embryo transfer avails for a separate cycle to focus on optimizing the chances of successful implantation. Children born from vitrified blastocysts have significantly higher birthweight than those born from non-frozen blastocysts. For early cleavage embryos, frozen ones appear to have at least as good obstetric outcome, measured as preterm birth and low birthweight for children born after cryopreservation as compared with children born after fresh cycles.

Oocyte age, survival proportion, and number of transferred embryos are predictors of pregnancy outcome.

Pregnancies have been reported from embryos stored for 16 years. A study of more than 11,000 cryopreserved human embryos showed no significant effect of storage time on post-thaw survival for IVF or oocyte donation cycles, or for embryos frozen at the pronuclear or cleavage stages. In addition, the duration of storage had no significant effect on clinical pregnancy, miscarriage, implantation, or live birth rate, whether from IVF or oocyte donation cycles.

A study in France between 1999 and 2011 came to the result that embryo freezing before administration of gonadotoxic chemotherapy agents to females caused a delay of treatment in 34% of cases, and a live birth in 27% of surviving cases who wanted to become pregnant, with the follow-up time varying between 1 and 13 years.

Legislation

From 1 October 2009 human embryos are allowed to be stored for 10 years in the UK, according to the Human Fertilisation and Embryology Act 2008.

History

The first ever pregnancy derived from a frozen human embryo was reported by Alan Trounson & Linda Mohr in 1983 (although the fetus aborted spontaneously at ten weeks of gestation); the first term pregnancy derived from a frozen embryo was born in 1984. Since then and up to 2008 it is estimated that between 350,000 and half a million IVF babies have been born from embryos frozen at a controlled rate and then stored in liquid nitrogen; additionally a few hundred births have been born from vitrified oocytes but firm figures are hard to come by. It may be noted that Subash Mukhopadyay from Kolkata, India reported the successful cryopreservation of an eight cell embryo, storing it for 53 days, thawing and replacing it into the mother's womb, resulting in a successful and live birth as early as 1978- a full five years before Trounson and Mohr had done so. A small publication of Mukherjee in 1978 clearly shows that Mukherjee was on the right line of thinking much before anyone else had demonstrated the successful outcome of a pregnancy following the transfer of a 8-cell frozen-thawed embryo into human subjects transferring 8-cell cryopreserved embryos." (Current Science, Vol .72. No. 7, 10 April 1997)

Cryoconservation of Animal Genetic Resources

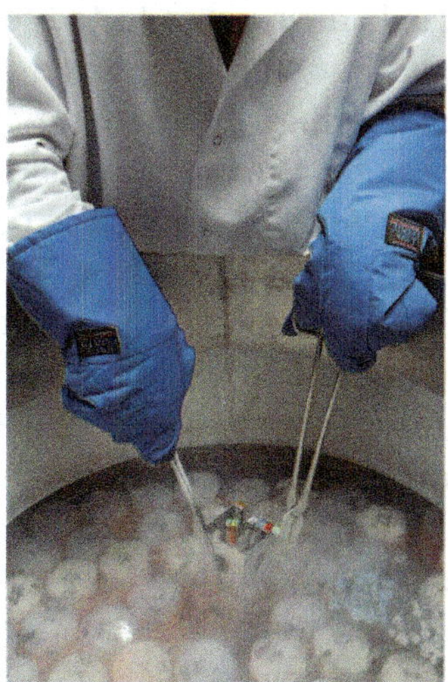

Cryoconservation of animal genetic resources at the USDA Gene Bank

Cryoconservation of animal genetic resources is a strategy wherein samples of animal genetic materials are preserved cryogenically. Animal genetic resources, as defined by the Food and Agriculture Organization of the United Nations, are "those animal species that are used, or may be used, for the production of food and agriculture, and the populations within each of them. These populations within each species can be classified as wild and feral populations, landraces and primary populations, standardised breeds, selected lines, varieties, strains and any conserved genetic material; all of which are currently categorized as Breeds." Genetic materials that are typically cryogenically preserved include sperm, oocytes, embryos and somatic cells. Cryogenic facilities are called gene banks and can vary greatly in size usually according to the economic resources available. They must be able to facilitate germplasm collection, processing, freezing, and long term storage, all in a hygienic and organized manner. Gene banks must maintain a precise database and make information and genetic resources accessible to properly facilitate cryoconservation. Cryoconservation is an *ex situ* conservation strategy that often coexists alongside *in situ* conservation to protect and preserve livestock genetics. Cryoconservation of livestock genetic resources is primarily done in order to preserve the genetics of populations of interest, such as indigenous breeds, also known as local or minor breeds. Material may be stored because individuals shared specific genes and phenotypes that may be of value or have potential value for researchers or breeders. Therefore, one of the main goals remains preserving the gene pool of local breeds that may be threatened. Indigenous livestock genetics are commonly threatened by factors such as globalization, modernization, changes in production systems, inappropriate introduction of major breeds, genetic drift, inbreeding, crossbreeding, climate change, natural disasters, disease, cultural changes, and urbanization. Indigenous livestock are critical to sustainable agricultural development and food security, due to their: adaptation to environment and endemic diseases, indispensable part in local production systems, social and cultural significance, and importance to local rural economies. The genetic resources of minor breeds have value to the local farmers, consumers of the products, private companies and investors interested in crossbreeding, breed associations, governments, those conducting research and development, and non-governmental organizations. Therefore, efforts have been made by national governments and non-governmental organizations, such as the Livestock Conservancy, to encourage conservation of livestock genetics through cryoconservation, as well as through other *ex situ* and *in situ* strategies. Cryogenic specimens of livestock genetic resources can be preserved and used for extended periods of time. This advantage makes cryoconservation beneficial particularly for threatened breeds who have low breed populations. Cryogenically preserved specimens can be used to revive breeds that are endangered or extinct, for breed improvement, crossbreeding, research and development. However, cryoconservation can be an expensive strategy and requires long term hygienic and economic commitment for germplasms to remain viable. Cryoconservation can also face unique challenges based on the species, as some species have a reduced survival rate of frozen germplasm.

Description

Cryoconservation is the process of freezing cells and tissues using liquid nitrogen to achieve extreme low temperatures with the intent of using the preserved sample to prevent the loss of genetic diversity. Semen, embryos, oocytes, somatic cells, nuclear DNA, and other types of biomaterial such as blood and serum can be stored using cryopreservation, in order to preserve genetic materials. The primary benefit of cryoconservation is the ability to save germplasms for extended periods of time, therefore maintaining the genetic diversity of a species or breed. There are two common techniques of cryopreservation: slow freezing and vitrification. Slow freezing helps eliminate the risk of intracellular ice crystals. If ice crystals form in the cells, there can be damage or destruction of genetic material. Vitrification is the process of freezing without the formation of ice crystals.

Value

Cryoconservation is an indispensable tool in the storage of genetic material of animal origin and will continue to be useful for the conservation of livestock into the future. Cryoconservation serves as a way to preserve germplasms, which is particularly beneficial for threatened breeds. Indigenous livestock may be conserved for a variety of reasons, including the preservation of local genetics, their importance in local traditions and their value to the culture identity and heritage of the area. The loss of regional livestock diversity could increase instability, decreases future possibilities and challenge production systems. Moreover, the maintenance of indigenous breeds can aid in the preservation of traditional lifestyles and livelihoods, even providing income through cultural tourism. Indigenous breeds can contribute to local economies and production systems by utilising land that is unsuitable for crop production to produce food products, as well as providing hides, manure and draft power. Therefore, the conservation and progression of these breeds are of the utmost importance for food security and sustainability.

Another beneficial factor in cryoconservation of indigenous livestock is in terms of food security and economic development. Indigenous livestock often have beneficial traits related to adaptation to local climate and diseases that can be incorporated into major breeds through cryoconservation practices. Cryoconservation is a favorable strategy because it allows germplasms to be stored for extended periods of time in a small confined area. An additional benefit of cryoconservation is the ability to preserve the biological material of both maternal and paternal cells and maintain viability over extended periods of time. Cryoconservation has been successfully used as a conservation strategy for species and breeds that have since been endangered. One drawback is that cryoconservation can only be done if preparation has taken place in advance. With proper preparation of collecting and maintaining genetic material, this method is very beneficial for the conservation of rare and endangered livestock. Cryoconservation can serve as a contingency plan when a breed population needs to be restored or when a breed has become extinct, as well as for breed improvement. This process benefits companies and researchers by making genetic materials available.

Conservation Goals

Flexibility of country's AGR to meet changes	Insurance against changes in production conditions	Safeguarding against diseases, disasters, etc.	Opportunities for genomic research

Genetic Factors	Allowing continued breed evolution/genetic adaption	Increasing knowledge of phenotypic characteristics of breed	Minimizing exposed to genetic drafts
Sustainable utilization of total areas	Opportunities for development in rural areas	Maintenance of agro-ecosystem diversity	Conservation of rural culture diversity

The support of numerous stakeholders make this process possible in the establishment and operations of cryoconservation. Before every phase is executed, all participating stakeholders must be briefed to understand the possible phase impending. This would include informing the stakeholders of their responsibilities and receiving their consent for the cryoconservation process. The possible stakeholders within the cryoconservation process could include:

- The State-the government acquires responsibility for conservation of animal genetic resources
- Individual Livestock Keepers and Breed Associations-individual livestock keepers are commonly the primary owners of the livestock whose germplasm is used for processes of cryoconservation. Breed Associations would be interested in the well-being of their respective breeds in short and long terms. Through this interest these associations may provide financial and organizational support for the cryoconservation process.
- Private Companies-including, but, not limited to, commercial breeding companies, processing companies and agricultural support services may find value in the cryoconservation process and may striving to become more involved.
- The National Coordinator for the Management of Animal Genetic Resources-this particular stakeholder would possibly a member of the National Advisory Committee on Animal Genetic Resources. This member needs to be knowledgeable about all aspects and activities of cryoconservation, as this stakeholder would have the responsibility of reporting current information to the FAO.

Methods

Collection

There are several ways to collect the genetic materials based on which type of germplasm.

Semen

Freezing semen is a commonly used technique in the modern animal agriculture industry, which is well researched with established methods Semen is often collected using an artificial vagina, electroejaculation, gloved-hand technique, abdominal stroking, or epididymal sperm collection. Preferred collection techniques vary based on species and available tools. Patience and technique are keys to successful collection of semen. There are several styles and types of artificial vaginas that can be used depending on the breed and species of the male. During this process the penis enters a tube that is the approximate pressure and temperature of the female's vagina. There is a disposable bag inside the tube that collects the semen. During this process it may be beneficial to have a teaser animal—an animal used to sexually tease but not impregnate the animal—to increase the arousal of the male. Electroejaculation is a method of semen collection in the cattle industry because it yields high quality semen. However, this process requires the animal to be trained and securely held, thus it is not ideal when working with wild or feral animals. When performing this process the electroejaculator is inserted into the rectum of the male. The electroejaculator stimulates the male causing an ejaculation, after which the semen is collected. The glove hand collection technique is used mainly in the swine industry. During this process, the boar mounts a dummy, while the handler grasps the penis of the boar between the ridges of his fingers and collects the semen. Abdominal stroking is exclusively used in the poultry industry. During the technique, one technician will hold the bird, while a second technician massages the bird's cloaca. However, feces and semen both exit the male bird's body through the cloaca, so the semen quality is often low.

Embryo

Embryo collection is more demanding and requires more training than semen collection because the female reproductive organs are located inside of the body cavity. Superovulation is a technique used in order to have a female release more oocytes than normal. This can be achieved by using hormones to manipulate the female's reproductive organs. The hormones used are typically gonadotropin-like, meaning they stimulate the gonads. Follicle stimulating hormone is the preferred hormone in cattle, sheep and goats. While in pigs, equine chorionic gonadotropin is preferred. However, this is not commonly done in the swine industry because gilts and sows (female pigs) naturally ovulate more than one oocyte at one time. Superovulation can be difficult because not all females will respond the same way and success will vary by species. Once the female has released the oocytes, they are fertilized internally—in vivo—and flushed out of her body. In vivo fertilization is more successful than in vitro fertilization. In cattle, usually 10 or more embryos are removed from the flushing process. In order to flush the uterus, a technician will first seal off the female's cervix and add fluid, which allows the ovum to be flushed out of the uterine horns and into a cylinder for analysis. This process typically takes 30 minutes or less. Technicians are able to determine the sex of the embryo, which can be especially beneficial in the dairy industry because it is more desirable for the embryo to be a female. Vitrification is the preferred method of embryo freezing because it yields higher quality embryos. It is crucial technicians handle the embryos with care and freeze them within 3–4 hours in order to preserve viability of the greatest percentage of embryos.

Oocytes

Oocytes can be collected from most mammalian species. Conventional oocyte collection is when ovaries are removed from a donor animal; this is done posthumously in slaughter facilities. The ovaries are kept warm as they are brought back to a laboratory for oocyte collection. Keeping the ovaries warm helps increase the success rate of fertilization. Once collected the oocytes are assessed and categorized into small, medium, and large, and then matured for 20–23 hours. This simple, inexpensive technique can lead to about 24 oocytes collected from a bovine. Conventional oocyte collection is especially useful for females who unexpectedly die or who are incapable of being bred due to injury. A second option for oocyte collection is to utilize the transvaginal ultrasound guided oocyte collection method otherwise known as TUGA. Collection technique varies slightly by species, but the general methods for collection are the same; a needle is inserted into each ovarian follicle and pulled out via vacuum. The major benefit of using this method is the ability to expand the lifetime reproductive productivity, or the number of productive days an animal is in her estrous cycle. Pregnant cows and mares continue to develop new follicles until the middle of pregnancy. Thus, TUGA can be used to substantially increase the fitness of an individual because the female then has the potential produce more than one offspring per gestation.

Somatic Cells

Somatic cells are an additional resource which can be retrieved for gene banking, particularly in the cases of emergency wherein gametes cannot be collected or stored. Tissues can be taken from living animals or shortly after death. These tissues can be saved via cryopreservation or dehydrated. Blood cells can also be useful for DNA analysis such as comparing homozygosity It is recommended by the FAO that two vials of blood be drawn to reduce the chance that all samples will be lost from a particular animal. DNA can be extracted using commercial kits, making this an affordable and accessible strategy for collecting germplasms.

	Semen	**Semen and Oocytes**	**Embryos**
Number of samples needed to restore a breed	2000	100 of each	200
Backcrossing needed?	Yes	No	No
Mitochondrial genes included?	No	Yes	Yes
Collection Possible in livestock species	Mostly, not always	Yes, in some species. Operational for bovines	Yes, in some species. Operational for bovines
Cost of collection	$$	$$	$$$$
Cryopreservation possible?	Yes	Still in experimental stage	Operational in bovines, horses and sheep. Promising in pigs. Impossible in poultry

Utilization	Surgical or non-surgical insemination backcrossing for 4 generations	In vitro maturation/IVF followed by surgical or non-surgical ET	Surgical or non-surgical ET
Current feasibility	High	Medium	High depending on available resources

Freezing

There are two cryopreservation freezing methods: slow freezing and vitrification.

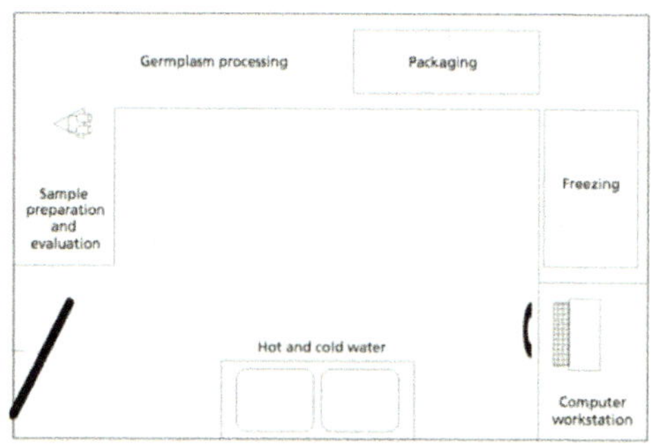

Example freezing laboratory

Slow Freezing

During slow freezing, cells are placed in a medium which is cooled below the freezing point using liquid nitrogen. This causes an ice mass to form in the medium. As the water in the medium freezes, the concentration of the sugars, salts, and cryoprotectant increase. Due to osmosis, the water from the cells enters the medium to keep the concentrations of sugars, salts, and cryoprotectant equal. The water that leaves the cells is eventually frozen, causing more water to diffuse out of the cell. Eventually, the unfrozen portion—cellular—becomes too viscous for ice crystals to form inside of the cell.

Vitrification

The second technique for cryoconservation is vitrification or flash freezing. Vitrification is the transformation from a liquid to solid state without the formation of crystals. The process and mechanics of vitrification are similar to slow freezing, the difference lying in the concentration of the medium. The vitrification method applies a selected medium which has a higher concentration of solute so the water will leave the cells via osmosis. The medium is concentrated enough so all of the intracellular water will leave without the medium needing to be reconcentrated. The higher concentration of the medium in vitrification allows the germplasms to be frozen more rapidly than with slow freezing. Vitrification is considered to be the more effective technique of freezing germplasms.

Facility Design and Equipment

Facility Design

When designing a facility, there are several things that should be kept in mind including biosecurity, worker safety and efficiency, and animal welfare. Diverse infrastructure is required in order to successfully collect and store genetic material. The buildings needed depend on the size of facilities as well as the extent of the operations.

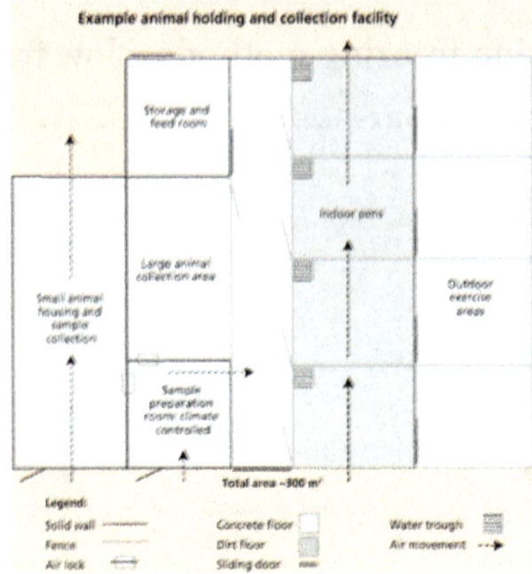

Example of animal holding and collecting facility

Biosecurity

Biosecurity, a management measure used to prevent the transmission of diseases and disease agents on the facility, is important to keep in mind when designing a facility. In order to achieve a high level of biosecurity, collection facilities should be placed as far as possible from one another, as well as from farms. According to the FAO's recommendations, facilities should be "at least 3 km from farms or other biological risks and 1 km from main roads and railways". Separation between collection facilities and surrounding farms can improve biosecurity as pests, such as flies and mice, have the potential to travel from farm to facility and vice versa. Other disease agents may be able to travel through the air via wind, furthering the importance of separation of farms and proper air sanitation and ventilation. Additionally, a perimeter fence is used to prevent potential threats that could cause contamination to germplasms, such as unauthorized personnel or unwanted animals, from entering the facilities. Animals may be housed in pens located inside or outside of a barn as long as they are contained within the perimeter fence. When interaction with outside objects, such as feed trucks or veterinary personnel, is necessary, complete sanitation is required to decrease the risk of contamination. There is always the possibility of disease spreading among the animals whose biological data is being collected or from animal to human. An example of a disease that can easily spread through germplasm is Porcine Reproductive and Respiratory Syndrome, otherwise known as PRRS. A highly contagious disease between swine, PRRS causes millions of dollars to be lost annually by producers. The disease can be spread through boar semen. Therefore, biosecurity is particularly important when genetic material will be inserted into another animal to prevent the spread of such diseases.

Human Considerations

Worker safety is always a priority when handling livestock. Escape routes and alternative access throughout the facility are crucial for both the handlers and livestock. Germplasm storage and collection sites must include locker rooms for staff, which provide lockers, showers, and storage of clothing and footwear, in order to meet sanitation requirements.

Animal Considerations

Animal housing practical when collecting germplasms because they keep donor animals in an easily accessible area, making the process of collecting germplasms easier and more efficient. The species and breeds of animals housed should be considered while planning the facility; facilities should be big enough to meet animal welfare standards, yet small enough to reduce human contact and increase ease of handling while reducing stress of the animal. As the process of collecting germplasm may take several days, the animal may become stressed causing a lower quality of genetic material to be obtained. Thus, training the animal to become familiar with the process is key. Holding facilities for animals may also serve as a quarantine. Quarantine facilities are necessary in order to prevent the transmission of disease from animal to animal, animal to germplasm, germplasm to germplasm, and germplasm to animal. Introducing quarantine to separate the diseased animal(s) from the healthy should be done immediately. However, a quarantine does not always prevent the spread of disease.

Temperature Control and Ventilation

Temperature control and ventilation should be included in the design of the holding and collection facilities to keep the animals comfortable and healthy, while limiting stress during the germplasm collection process. Ventilation serves as an effective way to keep clean airflow throughout the facilities and eliminate odors Temperature control helps regulate the air quality and humidity level inside the barn.

Equipment

A freezing and processing laboratory for genetic materials can be on the same site as the holding and collecting facility. However, the laboratory must have higher sanitation standards. According to the FAO, a proper germplasm laboratory should include the following.

- Washable work surfaces, floors (non-slip) and walls

- Sufficient lighting and ventilation
- Hot and cold, purified water
- Electrical sockets
- Adequate storage for consumable materials.

Cryopreservation requires equipment to collect biological material and test tubes for storage. Price is highly variable based on the quality of the collection and storage materials. The life expectancy of tools should be considered when determining costs. In addition to traditional laboratory equipment, the FAO also suggests the following:

- Disposable coveralls
- Portable incubator
- Haemocytometer
- Semen straws and filling/sealing equipment
- Liquid nitrogen storage tank
- Liquid nitrogen
- Liquid nitrogen dry-shipper
- Equipment for determining sperm concentration (one or more of the following three):
- Spectrophotometer (fixed or portable)
- Makler counter chamber (or disposable counting chamber)
- Haemocytometer
- Straw filling and sealing equipment
- Freezing equipment (manual or programmable)
- Carbon dioxide incubator (for embryos)
- Laminar flow benches (for embryos)
- Dry liquid nitrogen shipping tanks
- Long-term liquid nitrogen storage tanks.

Limitations

Cryoconservation is limited by the cells and tissues that can be frozen and successfully thawed. Cells and tissues that can be successfully frozen are limited by their surface area. To keep cells and tissues viable, they must be frozen quickly to prevent ice crystal formation. Thus, a large surface area is beneficial. Another limitation is the species being preserved. There have been difficulties using particular methods of cryoconservation with certain species. For example, artificial insemination is more difficult in sheep than cattle, goats, pigs, or horses due to posterior folds in the cervix of ovines. Cryopreservation of embryos is dependent on the species and the stage of development of the embryo. Pig embryos are the most difficult to freeze, thaw, and utilize produce live offspring due to their sensitivity to chilling and high lipid content.

Legal Issues

The collection and utilization of genetic materials requires clear agreements between stakeholders with regards to their rights and responsibilities. The FAO and others, such as Mendelsohn, suggests that governments establish policies with regards to livestock genetic resources and their collection, storage, distribution, and utilization are governments. The FAO also recommends that national or regional livestock industries establish an advisory committee to advise and provide recommendations on policy. Livestock are traditionally a private good; in order to obtain ownership of genetic materials, gene banks have several strategies that they can deploy. Gene banks may either:

- buy the livestock in order to obtain and preserve genetic information
- have the germplasm donated by livestock owner
- pay a fee to the livestock owner for germplasm rights
- develop a contract with the livestock owner in order to obtain germplasm ownership only after set period of time, in order to prevent immediate acquisition of germplasm by competitors

One of the key elements of cryoconservation of livestock is open access to genetic materials, to make the resources of these conserved genetic materials accessible for utilization. Utilization should be based on sustainable use, development, and conservation, as well as improvement for the livestock industry. Government and non-governmental organizations recommend that genetic information should have open access for the following purposes:

- national public need
- non-research breeding by non-governmental organizations or private entities
- research for breed improvement, conservation of endangered breeds, and potential recovery of extinct breeds

Examples

Hungarian Grey Cattle

An example of the use of cryoconservation to prevent the extinction of a livestock breed is the case of the Hungarian Grey cattle, or Magya Szurke. Hungarian Grey cattle were once a dominant breed in southeastern Europe with a population of 4.9 million head in 1884. They were mainly used for draft power and meat. However, the population had decreased to 280,000 head by the end of World War II and eventually reached the low population of 187 females and 6 males from 1965 to 1970. The breed's decreased use was due primarily to the mechanization of agriculture and the adoption of major breeds, which yield higher milk production. The Hungarian government launched a project to preserve the breed, as it possesses valuable traits, such as stamina, calving ease, disease resistance, and easy adaptation to a variety of climates. The government program included various conservation strategies, including the cryopreservation of semen and embryos. The Hungarian government's conservation effort brought the population up to 10,310 in 2012, which shows significant improvement using cryoconservation.

The Gaur

Gaur, also known as the Indian bison, is the heaviest and most powerful of all wild cattle native to South and Southeast Asia. It is indicated in field data that the population of mature animals is about 5,200–18,000. Male and female Gaur both have distinctive humps between the head and shoulders, a dorsal ridge, prominent horns, and a dewlap which extends to the front legs.The Gaur grows 60% faster than domestic cattle, meaning farmers meat can be harvested at a faster rate, making beef production two to three times more profitable. Gaur meat is preferred over other breeds' meat among local people. Another benefit of the bovine is that it has the ability to sweat and tolerates heat well.

The Gaur population experienced a drastic decline of about 90% between the 1960s and 1990s due to poaching, commercial hunting, shrinking habitat, and the spreading of disease. According to the International Union for Conservation of Nature's Red List, the Gaur is a vulnerable species due to its declining population in Southeast Asia. Although the global Gaur population has declined by 30% over the past 30 years, the Gaur has a relatively stable population in India, due to protective efforts such as cryoconservation. The American Association of Zoos and Aquariums, Integrated Conservation Research (ICR), and Advanced Cell Technology have made efforts to use cryopreserved specimens of the Gaur through artificial insemination, embryo transfer, and cloning, respectively. Hybridization with domestic cattle has been successfully achieved by ICR, in order to create higher yielding, heat resistant cattle.

Liquid Nitrogen

Liquid nitrogen is nitrogen in a liquid state at an extremely low temperature. It is a colorless clear liquid with a density of 0.807 g/ml at its boiling point (−195.79 °C (77 K; −320 °F)) and a dielectric constant of 1.43. Nitrogen was first liquefied at the Jagiellonian University on 15 April 1883 by Polish physicists, Zygmunt Wróblewski and Karol Olszewski. It is produced industrially by fractional distillation of liquid air. Liquid nitrogen is often referred to by the abbreviation, LN_2 or "LIN" or "LN" and has the UN number 1977. Liquid nitrogen is a diatomic liquid, which means that the diatomic character of the covalent N bonding in N_2 gas is retained after liquefaction.

Liquid nitrogen

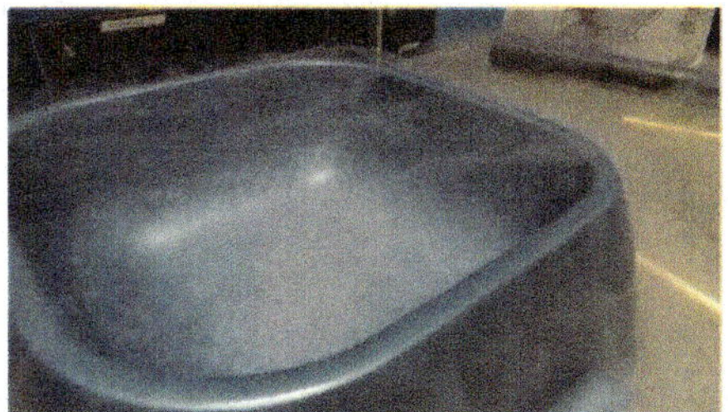

A demonstration of liquid nitrogen at the Freeside maker space in Atlanta, Georgia during the Online News Association conference in 2013

Students preparing homemade ice cream with liquid nitrogen.

Liquid nitrogen is a cryogenic fluid that can cause rapid freezing on contact with living tissue. When appropriately insulated from ambient heat, liquid nitrogen can be stored and transported, for example in vacuum flasks. The temperature is held constant at 77 K by slow boiling of the liquid, resulting in the evolution of nitrogen gas. Depending on the size and design, the holding time of vacuum flasks (Dewars) ranges from a few hours to a few weeks. The development of pressurised super-insulated vacuum vessels has enabled liquefied nitrogen to be stored and transported over longer time periods with losses reduced to 2% per day or less.

The temperature of liquid nitrogen can readily be reduced to its freezing point 63 K (−210 °C; −346 °F) by placing it in a vacuum chamber pumped by a vacuum pump. Liquid nitrogen's efficiency as a coolant is limited by the fact that it boils immediately on contact with a warmer object, enveloping the object in insulating nitrogen gas. This effect, known as the Leidenfrost effect, applies to any liquid in contact with an object significantly hotter than its boiling point. Faster cooling may be obtained by plunging an object into a slush of liquid and solid nitrogen rather than liquid nitrogen alone.

Uses

Liquid nitrogen is a compact and readily transported source of dry nitrogen gas, as it does not require pressurization. Further, its ability to maintain temperatures far below the freezing point of water makes it extremely useful in a wide range of applications, primarily as an open-cycle refrigerant, including:

- in cryotherapy for removing unsightly or potentially malignant skin lesions such as warts and actinic keratosis
- to store cells at low temperature for laboratory work
- in cryogenics
- in a cryophorus to demonstrate rapid freezing by evaporation
- as a backup nitrogen source in hypoxic air fire prevention systems
- as a source of very dry nitrogen gas
- for the immersion, freezing, and transportation of food products
- for the cryopreservation of blood, reproductive cells (sperm and egg), and other biological samples and materials
    - to preserve tissue samples from surgical excisions for future studies
    - to facilitate Cryoconservation of animal genetic resources
- as a method of freezing water and oil pipes in order to work on them in situations where a valve is not available to block fluid flow to the work area, method known as a cryogenic isolation (frequently used in industry and New York district steam pipework) – Electrical heat pumps are now often used for small pipe diameters.
- in the process of promession, a way to dispose of the dead
- for cryonic preservation in hopes of future reanimation.
- to shrink-weld machinery parts together

- as a coolant
    - for CCD cameras in astronomy
    - for a high-temperature superconductor to a temperature sufficient to achieve superconductivity
    - for vacuum pump traps and in controlled-evaporation processes in chemistry.
    - to increase the sensitivity of infrared homing seeker heads of missiles such as the Strela 3
    - to temporarily shrink mechanical components during machine assembly and allow improved interference fits
    - for computers and extreme overclocking
    - for simulation of space background in vacuum chamber during spacecraft thermal testing
- in food preparation, such as for making ultra-smooth ice cream.
- in container inerting and pressurisation by injecting a controlled amount of liquid nitrogen just prior to sealing or capping.
- as a cosmetic novelty giving a smoky, bubbling "cauldron effect" to drinks. liquid nitrogen cocktail.
- as an energy storage medium.
- branding cattle.

Culinary Use of Liquid Nitrogen

The culinary use of liquid nitrogen is mentioned in an 1890 recipe book titled *Fancy Ices* authored by Mrs. Agnes Marshall, but has been employed in more recent times by restaurants in the preparation of frozen desserts, such as ice cream, which can be created within moments at the table because of the speed at which it cools food. The rapidity of chilling also leads to the formation of smaller ice crystals, which provides the dessert with a smoother texture. The technique is employed by chef Heston Blumenthal who has used it at his restaurant, The Fat Duck to create frozen dishes such as egg and bacon ice cream. Liquid nitrogen has also become popular in the preparation of cocktails because it can be used to quickly chill glasses or freeze ingredients. It is also added to drinks to create a smoky effect, which occurs as tiny droplets of the liquid nitrogen come into contact with the surrounding air, condensing the vapour that is naturally present.

Safety

Filling a liquid nitrogen Dewar from a storage tank

Because the liquid-to-gas expansion ratio of nitrogen is 1:694 at 20 °C (68 °F), a tremendous amount of force can be generated if liquid nitrogen is rapidly vaporized in an enclosed space. In an incident on January 12, 2006 at Texas A&M University, the pressure-relief devices of a tank of liquid nitrogen were malfunctioning and later sealed. As a result of the subsequent pressure buildup, the tank failed catastrophically. The force of the explosion was sufficient to propel the tank through the ceiling immediately above it, shatter a reinforced concrete beam immediately below it, and blow the walls of the laboratory 0.1–0.2 m off their foundations.

Because of its extremely low temperature, careless handling of liquid nitrogen and any objects cooled by it may result in cold burns. In that case, special gloves should be used while handling. However, a small splash or even pouring down skin will not burn immediately, because the evaporating gas thermally insulates to some extent, like touching a hot element very briefly with a wet finger. If the liquid nitrogen pools anywhere, it will burn severely.

As liquid nitrogen evaporates it reduces the oxygen concentration in the air and can act as an asphyxiant, especially in confined spaces. Nitrogen is odorless, colorless, and tasteless and may produce asphyxia without any sensation or prior warning.

Oxygen sensors are sometimes used as a safety precaution when working with liquid nitrogen to alert workers of gas spills into a confined space.

Vessels containing liquid nitrogen can condense oxygen from air. The liquid in such a vessel becomes increasingly enriched in oxygen (boiling point 90 K; −183 °C; −298 °F) as the nitrogen evaporates, and can cause violent oxidation of organic material.

Ingestion of liquid nitrogen can cause severe internal damage. For example, in 2012, a young woman in England had her stomach removed after ingesting a cocktail made with liquid nitrogen.

Production

Liquid nitrogen is produced commercially from the cryogenic distillation of liquified air or from the liquefication of pure nitrogen derived from air using pressure swing adsorption. An air compressor is used to compress filtered air to high pressure; the high-pressure gas is cooled back to ambient temperature, and allowed to expand to a low pressure. The expanding air cools greatly (the Joule–Thomson effect), and oxygen, nitrogen, and argon are separated by further stages of expansion and distillation. Small-scale production of liquid nitrogen is easily achieved using this principle. Liquid nitrogen may be produced for direct sale, or as a byproduct of manufacture of liquid oxygen used for industrial processes such as steelmaking. Liquid-air plants producing on the order of tons per day of product started to be built in the 1930s but became very common after the Second World War; a large modern plant may produce 3000 tons/day of liquid air products.

Glass Transition

The glass–liquid transition or glass transition for short is the reversible transition in amorphous materials (or in amorphous regions within semicrystalline materials) from a hard and relatively brittle "glassy" state into a viscous or rubbery state as the temperature is increased. An amorphous solid that exhibits a glass transition is called a glass. The reverse transition, achieved by supercooling a viscous liquid into the glass state, is called vitrification.

The glass-transition temperature T_g of a material characterizes the range of temperatures over which this glass transition occurs. It is always lower than the melting temperature, T_m, of the crystalline state of the material, if one exists.

Hard plastics like polystyrene and poly(methyl methacrylate) are used well below their glass transition temperatures, that is in their glassy state. Their T_g values are well above room temperature, both at around 100 °C (212 °F). Rubber elastomers like polyisoprene and polyisobutylene are used above their T_g, that is, in the rubbery state, where they are soft and flexible.

Despite the change in the physical properties of a material through its glass transition, the transition is not considered a phase transition; rather it is a phenomenon extending over a range of temperature and defined by one of several conventions. Such conventions include a constant cooling rate (20 kelvins per minute (36 °F/min)) and a viscosity threshold of 10^{12} Pa·s, among others. Upon cooling or heating through this glass-transition range, the material also exhibits a smooth step in the thermal-expansion coefficient and in the specific heat, with the location of these effects again being dependent on the history of the material. The question of whether some phase transition underlies the glass transition is a matter of continuing research.

Introduction

The glass transition of a liquid to a solid-like state may occur with either cooling or compression. The transition comprises a smooth increase in the viscosity of a material by as much as 17 orders of magnitude without any pronounced change in material structure. The consequence of this dramatic increase is a glass exhibiting solid-like mechanical properties on the timescale of practical observation. This transition is in contrast to the freezing or crystallization transition, which is a first-order phase transition in the Ehrenfest classification and involves discontinuities in thermodynamic and dynamic properties such as volume, energy, and viscosity. In many materials that normally undergo a freezing transition, rapid cooling will avoid this phase transition and instead result in a glass transition at some lower temperature. Other materials, such as many polymers, lack a well defined crystalline state and easily form glasses, even upon very slow cooling or compression. The tendency for a material to form a glass while quenched is called glass forming ability. This ability depends on the composition of the material and can be predicted by the rigidity theory.

Below the transition temperature range, the glassy structure does not relax in accordance with the cooling rate used. The expansion coefficient for the glassy state is roughly equivalent to that of the crystalline solid. If slower cooling rates are used, the increased time for structural relaxation (or intermolecular rearrangement) to occur may result in a higher density glass product. Similarly, by annealing (and thus allowing for slow structural relaxation) the glass structure in time approaches an equilibrium density corresponding to the supercooled liquid at this same temperature. T_g is located at the intersection between the cooling curve (volume versus temperature) for the glassy state and the supercooled liquid.

The configuration of the glass in this temperature range changes slowly with time towards the equilibrium structure. The principle of the minimization of the Gibbs free energy provides the thermodynamic driving force necessary for the eventual change. It should be noted here that at somewhat higher temperatures than T_g, the structure corresponding to equilibrium at any temperature is achieved quite rapidly. In contrast, at considerably lower temperatures, the configuration of the glass remains sensibly stable over increasingly extended periods of time.

Thus, the liquid-glass transition is not a transition between states of thermodynamic equilibrium. It is widely believed that the true equilibrium state is always crystalline. Glass is believed to exist in a kinetically locked state, and its entropy, density, and so on, depend on the thermal history. Therefore, the glass transition is primarily a dynamic phenomenon. Time and temperature are interchangeable quantities (to some extent) when dealing with glasses, a fact often expressed in the time–temperature superposition principle. On cooling a liquid, *internal degrees of freedom successively fall out of equilibrium*. However, there is a longstanding debate whether there is an underlying second-order phase transition in the hypothetical limit of infinitely long relaxation times.

Transition Temperature T_g

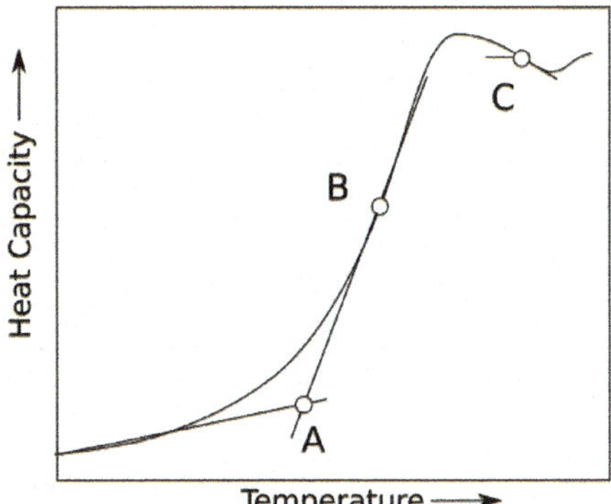

Measurement of T_g (the temperature at the point A) by differential scanning calorimetry

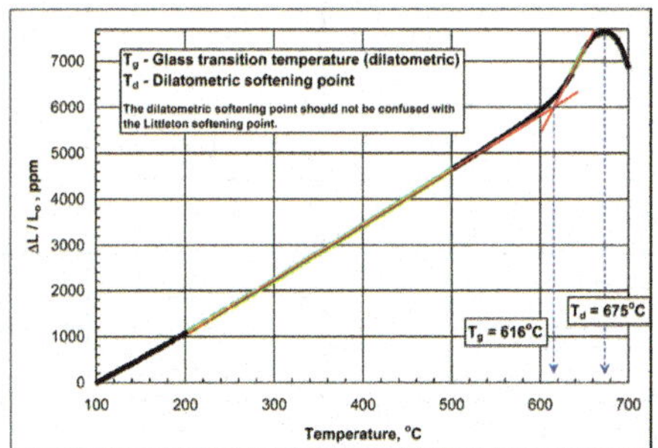

Determination of T_g by dilatometry.

Refer to the figure on the right plotting the heat capacity as a function of temperature. In this context, T_g is the temperature corresponding to point A on the curve. The linear sections below and above T_g are colored green. T_g is the temperature at the intersection of the red regression lines.

Different operational definitions of the glass transition temperature T_g are in use, and several of them are endorsed as accepted scientific standards. Nevertheless, all definitions are arbitrary, and all yield different numeric results: at best, values of T_g for a given substance agree within a few kelvins. One definition refers to the viscosity, fixing T_g at a value of 10^{13} poise (or 10^{12} Pa·s). As evidenced experimentally, this value is close to the annealing point of many glasses.

In contrast to viscosity, the thermal expansion, heat capacity, shear modulus, and many other properties of inorganic glasses show a relatively sudden change at the glass transition temperature. Any such step or kink can be used to define T_g. To make this definition reproducible, the cooling or heating rate must be specified.

The most frequently used definition of T_g uses the energy release on heating in differential scanning calorimetry. Typically, the sample is first cooled with 10 K/min and then heated with that same speed.

Yet another definition of T_g uses the kink in dilatometry (a.k.a. thermal expansion). Here, heating rates of 3–5 K/min (5.4–9.0 °F/min) are common. Summarized below are T_g values characteristic of certain classes of materials.

Polymers

Material	T_g (°C)	T_g (°F)	Commercial name
Tire rubber	−70	−94	
Polyvinylidene fluoride (PVDF)	−35	−31	
Polypropylene (PP atactic)	−20	−4	
Polyvinyl fluoride (PVF)	−20	−4	
Polypropylene (PP isotactic)	0	32	
Poly-3-hydroxybutyrate (PHB)	15	59	
Poly(vinyl acetate) (PVAc)	30	86	
Polychlorotrifluoroethylene (PCTFE)	45	113	
Polyamide (PA)	47–60	117–140	Nylon-6,x
Polylactic acid (PLA)	60–65	140–149	
Polyethylene terephthalate (PET)	70	158	
Poly(vinyl chloride) (PVC)	80	176	
Poly(vinyl alcohol) (PVA)	85	185	
Polystyrene (PS)	95	203	
Poly(methyl methacrylate) (PMMA atactic)	105	221	Plexiglas
Acrylonitrile butadiene styrene (ABS)	105	221	
Polytetrafluoroethylene (PTFE)	115	239	Teflon
Poly(carbonate) (PC)	145	293	Lexan
Polysulfone	185	365	
Polynorbornene	215	419	

Dry nylon-6 has a glass transition temperature of 47 °C (117 °F). Nylon-6,6 in the dry state has a glass transition temperature of about 70 °C (158 °F). Whereas polyethene has a glass transition range of −130 – −80 °C (−202 – −112 °F) The above are only mean values, as the glass transition temperature depends on the cooling rate and molecular weight distribution and could be influenced by additives. For a semi-crystalline material, such as polyethene that is 60–80% crystalline at room temperature, the quoted glass transition refers to what happens to the amorphous part of the material upon cooling.

Silicates and Other Covalent Network Glasses

Material	T_g (°C)	T_g (°F)
Chalcogenide GeSbTe	150	302
Chalcogenide AsGeSeTe	245	473
ZBLAN fluoride glass	235	455
Tellurium dioxide	280	536
Fluoroaluminate	400	752
Soda-lime glass	520–600	968–1,112
Fused quartz (approximate)	1,200	2,200

Kauzmann's Paradox

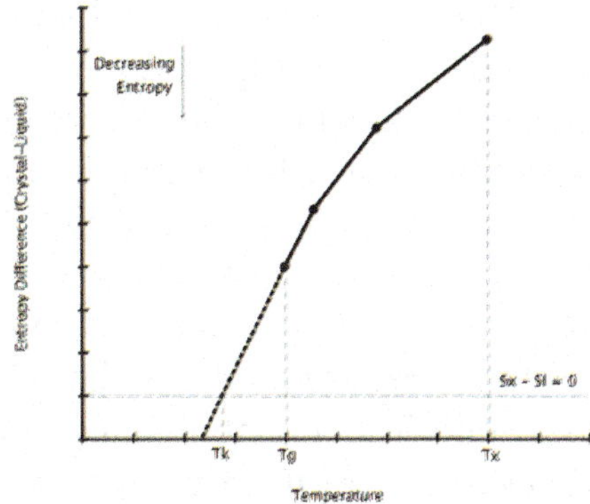

Entropy difference between crystal and undercooled melt

As a liquid is supercooled, the difference in entropy between the liquid and solid phase decreases. By extrapolating the heat capacity of the supercooled liquid below its glass transition temperature, it is possible to calculate the temperature at which the difference in entropies becomes zero. This temperature has been named the Kauzmann temperature.

If a liquid could be supercooled below its Kauzmann temperature, and it did indeed display a lower entropy than the crystal phase, the consequences would be paradoxical. This Kauzmann paradox has been the subject of much debate and many publications since it was first put forward by Walter Kauzmann in 1948.

One resolution of the Kauzmann paradox is to say that there must be a phase transition before the entropy of the liquid decreases. In this scenario, the transition temperature is known as the *calorimetric ideal glass transition temperature* T_{oc}. In this view, the glass transition is not merely a kinetic effect, i.e. merely the result of fast cooling of a melt, but there is an underlying thermodynamic basis for glass formation. The glass transition temperature:

$$T_g \to T_{0c} \text{ as } \frac{dT}{dt} \to 0.$$

There are at least three other possible resolutions to the Kauzmann paradox. It could be that the heat capacity of the supercooled liquid near the Kauzmann temperature smoothly decreases to a smaller value. It could also be that a first order phase transition to another liquid state occurs before the Kauzmann temperature with the heat capacity of this new state being less than that obtained by extrapolation from higher temperature. Finally, Kauzmann himself resolved the entropy paradox by postulating that all supercooled liquids must crystallize before the Kauzmann temperature is reached.

The glass Transition in Specific Materials

Silica, SiO_2

Silica (the chemical compound SiO_2) has a number of distinct crystalline forms in addition to the quartz structure. Nearly all of the crystalline forms involve tetrahedral SiO_4 units linked together by *shared vertices* in different arrangements. Si-O bond lengths vary between the different crystal forms. For example, in α-quartz the bond length is 161 picometres (6.3×10^{-9} in), whereas in α-tridymite it ranges from 154–171 pm (6.1×10^{-9}–6.7×10^{-9} in). The Si-O-Si bond angle also varies from 140° in α-tridymite to 144° in α-quartz to 180° in β-tridymite. Any deviations from these standard parameters constitute microstructural differences or variations that represent an approach to an amorphous, vitreous or glassy solid. The transition temperature T_g in silicates is related to the energy required to break and re-form covalent bonds in an amorphous (or random network) lattice of covalent bonds. The T_g is clearly influenced by the chemistry of the glass. For example, addition of elements such as B, Na, K or Ca to a silica glass, which have a valency less than 4, helps in breaking up the network structure, thus reducing the T_g. Alternatively, P, which has a valency of 5, helps to reinforce an ordered lattice, and thus increases the T_g.

T_g is directly proportional to bond strength, e.g. it depends on quasi-equilibrium thermodynamic parameters of the bonds e.g. on the enthalpy H_d and entropy S_d of configurons – broken bonds: $T_g = H_d / [S_d + R\ln[(1-f_c)/ f_c]$ where R is the gas constant and f_c is the percolation threshold. For strong melts such as SiO_2 the percolation threshold in the above equation is the universal Scher-Zallen critical density in the 3-D space e.g. $f_c = 0.15$, however for fragile materials the percolation thresholds are material-dependent and $f_c << 1$. The enthalpy H_d and the entropy S_d of configurons – broken bonds can be found from available experimental data on viscosity.

In ironing, a fabric is heated through the glass-rubber transition.

Polymers

In polymers the glass transition temperature, T_g, is often expressed as the temperature at which the Gibbs free energy is such that the activation energy for the cooperative movement of 50 or so elements of the polymer is exceeded. This allows molecular chains to slide past each other when a force is applied. From this definition, we can that the introduction of relatively stiff chemical groups (such as benzene rings) will interfere with the flowing process and hence increase T_g. The stiffness of thermoplastics decreases due to this effect. When the glass temperature has been reached, the stiffness stays the same for a while, i.e., at or near E_2, until the temperature exceeds T_m, and the material melts. This region is called the rubber plateau.

In ironing, a fabric is heated through this transition so that the polymer chains become mobile. The weight of the iron then imposes a preferred orientation. T_g can be significantly decreased by addition of plasticizers into the polymer matrix. Smaller molecules of plasticizer embed themselves between the polymer chains, increasing the spacing and free volume, and allowing them to move past one another even at lower temperatures. The "new-car smell" is due to the initial outgassing of volatile small-molecule plasticizers (most commonly known as phthalates) used to modify interior plastics (e.g., dashboards) to keep them from cracking in the cold of winter weather. The addition of nonreactive side groups to a polymer can also make the chains stand off from one another, reducing T_g. If a plastic with some desirable properties has a T_g that is too high, it can sometimes be combined with another in a copolymer or composite material with a T_g below the temperature of intended use. Note that some plastics are used at high temperatures, e.g., in automobile engines, and others at low temperatures.

In viscoelastic materials, the presence of liquid-like behavior depends on the properties of and so varies with rate of applied load, i.e., how quickly a force is applied. The silicone toy Silly Putty behaves quite differently depending on the time rate of applying a force: pull slowly and it flows, acting as a heavily viscous liquid; hit it with a hammer and it shatters, acting as a glass.

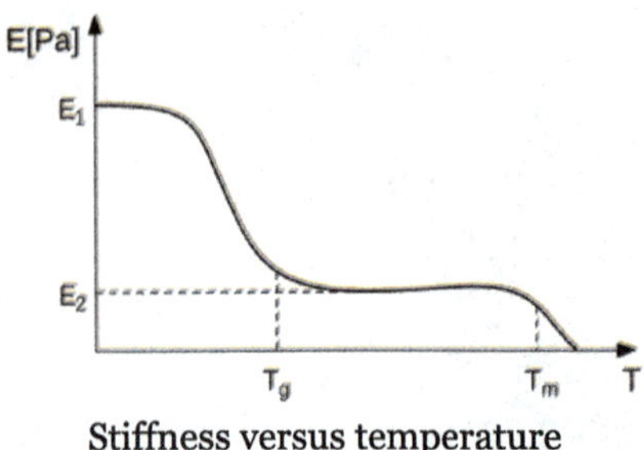

Stiffness versus temperature

On cooling, rubber undergoes a *liquid-glass transition*, which has also been called a *rubber-glass transition.*

Mechanics of Vitrification

Molecular motion in condensed matter can be represented by a Fourier series whose physical interpretation consists of a superposition of longitudinal and transverse waves of atomic displacement with varying directions and wavelengths. In monatomic systems, these waves are called *density fluctuations.* (In polyatomic systems, they may also include compositional fluctuations.)

Thus, thermal motion in liquids can be decomposed into elementary longitudinal vibrations (or acoustic phonons) while transverse vibrations (or shear waves) were originally described only in elastic solids exhibiting the highly ordered crystalline state of matter. In other words, simple liquids cannot support an applied force in the form of a shearing stress, and will yield mechanically via macroscopic plastic deformation (or viscous flow). Furthermore, the fact that a solid deforms locally while retaining its rigidity – while a liquid yields to macroscopic viscous flow in response to the application of an applied shearing force – is accepted by many as the mechanical distinction between the two.

The inadequacies of this conclusion, however, were pointed out by Frenkel in his revision of the kinetic theory of solids and the theory of elasticity in liquids. This revision follows directly from the continuous characteristic of the structural transition from the liquid state into the solid one when this transition is not accompanied by crystalliza-tion—ergo the supercooled viscous liquid. Thus we the intimate correlation between transverse acoustic phonons (or shear waves) and the onset of rigidity upon vitrifica-tion, as described by Bartenev in his mechanical description of the vitrification process.

The velocities of longitudinal acoustic phonons in condensed matter are directly responsible for the thermal conductivity that levels out temperature differentials between compressed and expanded volume elements. Kittel proposed that the behavior of glasses is interpreted in terms of an approximately constant "mean free path" for lattice phonons, and that the value of the mean free path is of the order of magnitude of the scale of disorder in the molecular structure of a liquid or solid. The thermal phonon mean free paths or relaxation lengths of a number of glass formers have been plotted versus the glass transition temperature, indicating a linear relationship between the two. This has suggested a new criterion for glass formation based on the value of the phonon mean free path.

It has often been suggested that heat transport in dielectric solids occurs through elastic vibrations of the lattice, and that this transport is limited by elastic scattering of acoustic phonons by lattice defects (e.g. randomly spaced vacancies). These predictions were confirmed by experiments on commercial glasses and glass ceramics, where mean free paths were apparently limited by "internal boundary scattering" to length scales of 10–100 micrometres (0.00039–0.00394 in). The relationship between these transverse waves and the mechanism of vitrification has been described by several authors who proposed that the onset of correlations between such phonons results in an orientational ordering or "freezing" of local shear stresses in glass-forming liquids, thus yielding the glass transition.

Electronic Structure

The influence of thermal phonons and their interaction with electronic structure is a topic that was appropriately introduced in a discussion of the resistance of liquid metals. Lindemann's theory of melting is referenced, and it is suggested that the drop in conductivity in going from the crystalline to the liquid state is due to the increased scattering of conduction electrons as a result of the increased amplitude of atomic vibration. Such theories of localization have been applied to transport in metallic glasses, where the mean free path of the electrons is very small (on the order of the interatomic spacing).

The formation of a non-crystalline form of a gold-silicon alloy by the method of splat quenching from the melt led to further considerations of the influence of electronic structure on glass forming ability, based on the properties of the metallic bond.

Other work indicates that the mobility of localized electrons is enhanced by the presence of dynamic phonon modes. One claim against such a model is that if chemical bonds are important, the nearly free electron models should not be applicable. However, if the model includes the buildup of a charge distribution between all pairs of atoms just like a chemical bond (e.g., silicon, when a band is just filled with electrons) then it should apply to solids.

Thus, if the electrical conductivity is low, the mean free path of the electrons is very short. The electrons will only be sensitive to the short-range order in the glass since they do not get a chance to scatter from atoms spaced at large distances. Since the short-range order is similar in glasses and crystals, the electronic energies should be similar in these two states. For alloys with lower resistivity and longer electronic mean free paths, the electrons could begin to sense that there is disorder in the glass, and this would raise their energies and destabilize the glass with respect to crystallization. Thus, the glass formation tendencies of certain alloys may therefore be due in part to the fact that the electron mean free paths are very short, so that only the short-range order is ever important for the energy of the electrons.

It has also been argued that glass formation in metallic systems is related to the "softness" of the interaction potential between unlike atoms. Some authors, emphasizing the strong similarities between the local structure of the glass and the corresponding crystal, suggest that chemical bonding helps to stabilize the amorphous structure.

Other authors have suggested that the electronic structure yields its influence on glass formation through the directional properties of bonds. Non-crystallinity is thus favored in elements with a large number of polymorphic forms and a high degree of bonding anisotropy. Crystallization becomes more unlikely as bonding anisotropy is increased from isotropic metallic to anisotropic metallic to covalent bonding, thus suggesting a relationship between the group number in the periodic table and the glass forming ability in elemental solids.

Ex Situ Conservation

Ex situ conservation literally means, "off-site conservation". It is the process of protecting an endangered species, variety or breed, of plant or animal outside of its natural habitat; for example, by removing part of the population from a threatened habitat and placing it in a new location, which may be a wild area or within the care of humans. The degree to which humans control or modify the natural dynamics of the managed population varies widely, and this may include alteration of living environments, reproductive patterns, access to resources, and protection from predation and mortality. *Ex situ* management can occur within or outside a species' natural geographic range. Individuals maintained *ex situ* exist outside of an ecological niche. This means that they are not under the same selection pressures as wild populations, and they may undergo artificial selection if maintained *ex situ* for multiple generations.

Agricultural biodiversity is also conserved in *ex situ* collections. This is primarily in the form of gene banks where samples are stored in order to conserve the genetic resources of major crop plants and their wild relatives.

Facilities

Botanical Gardens, Zoos, and Aquaria

Botanical gardens, and zoos are the most conventional methods of ex situ conservation, all of which house whole, protected specimens for breeding and reintroduction into the wild when necessary and possible. These facilities provide not only housing and care for specimens of endangered species, but also have an educational value. They inform the public of the threatened status of endangered species and of those factors which cause the threat, with the hope of creating public interest in stopping and reversing those factors which jeopardize a species' survival in the first place. They are the most publicly visited ex situ conservation sites, with the WZCS (World Zoo Conservation Strategy) estimating that the 1100 organized zoos in the world receive more than 600 million visitors annually. Globally there is an estimated total of 2,107 aquaria and zoos in 125 countries. Additionally many private collectors or other not-for-profit groups hold animals and they engage in conservation or reintroduction efforts. Similarly there are approximately 2,000 botanical gardens in 148 counties cultivating or storing an estimated 80,000 taxa of plants.

Techniques for Plants

Cryopreservation

The storage of seeds, pollen, tissue, or embryos in liquid nitrogen. This method can be used for virtually indefinite storage of material without deterioration over a much greater time-period relative to all other methods of *ex situ* conservation. Cryopreservation is also used for the conservation of livestock genetics through Cryoconservation of animal genetic resources. Technical limitations prevent the cryopreservation of many species, but cryobiology is a field of active research, and many studies concerning plants are underway.

Seed Banking

The storage of seeds in a temperature and moisture controlled environment. This technique is used for taxa with orthodox seeds that tolerate desiccation. Seed bank facilities vary from sealed boxes to climate controlled walk-in freezers or vaults. Taxa with recalcitrant seeds that do not tolerate desiccation are typically not held in seed banks for extended periods of time.

Tissue Culture (Storage and Propagation)

Somatic tissue can be stored for short periods of time in vitro for short periods of time. This is done in a light and temperature controlled environment that regulates the growth of cells. As a ex *situ* conservation technique tissue culture is primary used for clonal propagation of vegetative tissue or immature seeds. This allows for the proliferation of clonal plants from a relatively small amount of parent tissue.

Field Gene Banking

An extensive open-air planting used maintain genetic diversity of wild, agricultural, or forestry species. Typically species that are either difficult or impossible to conserve in seed banks are conserved in field gene banks. Field gene banks may also be used grow and select progeny of species stored by other *ex situ* techniques.

Cultivation Collections

Plants under horticultural care in a constructed landscape, typically a botanic garden or arboreta. This technique is similar to a field gene bank in that plants are maintained in the ambient environment, but the collections are typically not as genetically diverse or extensive. These collections are susceptible to hybridization, artificial selection, genetic drift, and disease transmission. Species that cannot be conserved by other *ex situ* techniques are often included in cultivated collections.

Inter Situ

Plants are under horticulture care, but the environment is managed to near natural conditions. This occurs with either restored or semi-natural environments. This technique is primarily used for taxa that are rare or in areas where habitat has been severely degraded.

Techniques for Animals

Endangered animal species and breeds are preserved using similar techniques. Animal species can be preserved in genebanks, which consist of cryogenic facilities used to store living sperm, eggs, or embryos. For example, the Zoological Society of San Diego has established a "frozen zoo" to store such samples using cryopreservation techniques from more than 355 species, including mammals, reptiles, and birds.

A potential technique for aiding in reproduction of endangered species is interspecific pregnancy, implanting embryos of an endangered species into the womb of a female of a related species, carrying it to term. It has been carried out for the Spanish ibex.

Genetic Management of Captive Populations

Captive populations are subject to problems such as inbreeding depression, loss of genetic diversity and adaptations to captivity. It is important to manage captive populations in a way that minimizes these issues so that the individuals to be introduced will resemble the original founders as closely as possible, which will increase the chances of successful reintroductions. During the initial growth phase, the population size is rapidly expanded until a target population size is reached. The target population size is the number of individuals that are required to maintain appropriate levels of genetic diversity, which is generally considered to by 90% of the current genetic diversity after 100 years. The number of individuals required to meet this goal varies based on potential growth rate, effective size, current genetic diversity, and generation time. Once the target population size is reached, the focus shifts to maintaining the population and avoiding genetic issues within the captive population.

Minimizing Mean Kinship

Managing populations based on minimizing mean kinship values is often an effective way to increase genetic diversity and to avoid inbreeding within captive populations. Kinship is the probability that two alleles will be identical by descent when one allele is taken randomly from each mating individual. The mean kinship value is the average kinship value between a given individual and every other member of the population. Mean kinship values can help determine which individuals should be mated. In choosing individuals for breeding, it is important to choose individuals with the lowest mean kinship values because these individuals are least related to the rest of the population and have the least common alleles. This ensures that rarer alleles are passed on, which helps to increase genetic diversity. It is also important to avoid mating two individuals with very different mean kinship values because such pairings propagate both the rare alleles that are present in the individual with the low mean kinship value as well as the common alleles that are present in the individual with the high mean kinship value. This genetic management technique requires that ancestry is known, so in circumstances where ancestry is unknown, it might be necessary to use molecular genetics such as microsatellite data to help resolve unknowns.

Avoiding Loss of Genetic Diversity

Genetic diversity is often lost within captive populations due to the founder effect and subsequent small population sizes. Minimizing the loss of genetic diversity within the captive population is an important component of *ex situ* conservation and is critical for successful reintroductions and the long term success of the species, since more diverse populations have higher adaptive potential. The loss of genetic diversity due to the founder effect can be minimized by ensuring that the founder population is large enough and genetically representative of the wild population. This is often difficult because removing large numbers of individuals from the wild populations may further reduce the genetic diversity of a species that is already of conservation concern. Maximizing the captive population size and the effective population size can decrease the loss of genetic diversity by minimizing the random loss of alleles due to genetic drift . Minimizing the number of generations in captivity is another effective method for reducing the loss of genetic diversity in captive populations.

Avoiding Adaptations to Captivity

Selection favors different traits in captive populations than it does in wild populations, so this may result in adaptations that are beneficial in captivity but are deleterious in the wild. This reduces the success of re-introductions, so it is important to manage captive populations in order to reduce adaptations to captivity. Adaptations to captivity can be reduced by minimizing the number of generations in captivity and by maximizing the number of migrants from wild populations. Minimizing selection on captive populations by creating an environment that is similar to their natural environment is another method of reducing adaptations to captivity, but it is important to find a balance between an environment that minimizes adaptation to captivity and an environment that permits adequate reproduction. Adaptations to captivity can also be reduced by managing the captive population as a series of population fragments. In this management strategy, the captive population is split into several sub-populations or fragments which are maintained separately. Smaller populations have lower adaptive potentials, so the population fragments are less likely to accumulate adaptations associated with captivity. The fragments are maintained separately until inbreeding becomes a concern. Immigrants are then exchanged between the fragments to reduce inbreeding, and then the fragments are managed separately again.

Managing Genetic Disorders

Genetic disorders are often an issue within captive populations due to the fact that the populations are usually established from a small number of founders. In large, outbreeding populations, the frequencies of most deleterious alleles are relatively low, but when a population undergoes a bottleneck during the founding of a captive population, previously rare alleles may survive and increase in number. Further inbreeding within the captive population may also increase the likelihood that deleterious alleles will be expressed due to increasing homozygosity within the population. The high occurrence of genetic disorders within a captive population can threaten both the survival of the captive population and its eventual reintroduction back into the wild. If the genetic disorder is dominant, it may be possible to eliminate the disease completely in a single generation by avoiding breeding of the affected individuals. However, if the genetic disorder is recessive, it may not be possible to completely eliminate the allele due to its presence in unaffected heterozygotes. In this case, the best option is to attempt to minimize the frequency of the allele by selectively choosing mating pairs. In the process of eliminating genetic disorders, it is important to consider that when certain individuals are prevented from breeding, alleles and therefore genetic diversity are removed from the population; if these alleles aren't present in other individuals, they may be lost completely. Preventing certain individuals from the breeding also reduces the effective population size, which is associated with problems such as the loss of genetic diversity and increased inbreeding.

Examples

Showy Indian clover, *Trifolium amoenum*, is an example of a species that was thought to be extinct, but was rediscovered in 1993 in the form of a single plant at a site in western Sonoma County. Seeds were harvested and currently grown in ex situ facilities.

The Wollemi pine is another example of a plant that is being preserved via *ex situ* conservation, as they are being grown in nurseries to be sold to the general public.

Drawbacks

Ex situ conservation, while helpful in humankind's efforts to sustain and protect our environment, is rarely enough to save a species from extinction. It is to be used as a last resort, or as a supplement to in situ conservation because it cannot recreate the habitat as a whole: the entire genetic variation of a species, its symbiotic counterparts, or those elements which, over time, might help a species adapt to its changing surroundings. Instead, *ex situ* conservation removes the species from its natural ecological contexts, preserving it under semi-isolated conditions whereby natural evolution and adaptation processes are either temporarily halted or altered by introducing the specimen to an unnatural habitat. In the case of cryogenic storage methods, the preserved specimen's adaptation processes are (quite literally) frozen altogether. The downside to this is that, when re-released, the species may lack the genetic adaptations and mutations which would allow it to thrive in its ever-changing natural habitat.

Furthermore, *ex situ* conservation techniques are often costly, with cryogenic storage being economically infeasible in most cases since species stored in this manner cannot provide a profit but instead slowly drain the financial resources of the government or organization determined to operate them. Seedbanks are ineffective for certain plant genera with recalcitrant seeds that do not remain fertile for long periods of time. Diseases and pests foreign to the species, to which the species has no natural defense, may also cripple crops of protected plants in *ex situ* plantations and in animals living in *ex situ* breeding grounds. These factors, combined with the specific environmental needs of many species, some of which are nearly impossible to recreate by man, make *ex situ* conservation impossible for a great number of the world's endangered flora and fauna.

Cryoprotectant

A cryoprotectant is a substance used to protect biological tissue from freezing damage (i.e. that due to ice formation). Arctic and Antarctic insects, fish and amphibians create cryoprotectants (antifreeze compounds and antifreeze proteins) in their bodies to minimize freezing damage during cold winter periods. Cryoprotectants are also used to preserve living materials in the study of biology and to preserve food products.

Mechanism

Cryoprotectants operate by increasing the solute concentration in cells. However, in order to be biologically viable they must easily penetrate cells and must not be toxic to cells.

Glass Transition Temperature

Some cryoprotectants function by lowering the glass transition temperature of a solution or of a material. In this way, the cryoprotectant prevents actual freezing, and the solution maintains some flexibility in a glassy phase. Many cryoprotectants also function by forming hydrogen bonds with biological molecules as water molecules are displaced. Hydrogen bonding in aqueous solutions is important for proper protein and DNA function. Thus, as the cryoprotectant replaces the water molecules, the biological material retains its native physiological structure and function, although they are no longer immersed in an aqueous environment. This preservation strategy is most often utilized in anhydrobiosis.

Toxicity

Mixtures of cryoprotectants have less toxicity and are more effective than single-agent cryoprotectants. A mixture of formamide with DMSO (dimethyl sulfoxide), propylene glycol, and a colloid was for many years, the most effective of all artificially created cryoprotectants. Cryoprotectant mixtures have been used for vitrification (i.e. solidification without crystal ice formation). Vitrification has important applications in preserving embryos, biological tissues, and organs for transplant. Vitrification is also used in cryonics in an effort to eliminate freezing damage.

Conventional

Conventional cryoprotectants are glycols (alcohols containing at least two hydroxyl groups), such as ethylene glycol, propylene glycol, and glycerol. Ethylene glycol is commonly used as automobile antifreeze, and propylene glycol has been used to reduce ice formation in ice cream. Dimethyl sulfoxide (DMSO) is also regarded as a conventional cryoprotectant. Glycerol and DMSO have been used for decades by cryobiologists to reduce ice formation in sperm, oocytes, and embryos that are cold-preserved in liquid nitrogen. Cryoconservation of animal genetic resources is a practice that involves conventional cryoprotectants to store genetic material with the intention of future revival. Trehalose is non-reducing sugar is produced by yeasts and insects in copious amounts. Its use as a cryoprotectant in commercial systems has been patented widely.

Examples in Nature

Insects

Insects most often use sugars or polyols as cryoprotectants. One species that uses cryoprotectant is *Polistes exclamans*. In this species, the different levels of cryoprotectant can be used to distinguish between morphologies.

Amphibians

Arctic frogs use glucose, but Arctic salamanders create glycerol in their livers for use as a cryoprotectant.

Food Preservation

Cryoprotectants are also used to preserve foods. These compounds are typically sugars that are inexpensive and do not pose any toxicity concerns. For example, many (raw) frozen chicken products contain a sucrose and sodium phosphates solution in water..

Common

- DMSO
- Ethylene glycol
- Glycerol
- 2-Methyl-2,4-pentanediol (MPD)
- Propylene glycol
- Sucrose
- Trehalose

Cryostasis (Clathrate Hydrates)

The term cryostasis was introduced to name the reversible preservation technology for live biological objects which is based on using clathrate-forming gaseous substances under increased hydrostatic pressure and hypothermic temperatures.

Living tissues cooled below the freezing point of water are damaged by the dehydration of the cells as ice is formed between the cells. The mechanism of freezing damage in living biological tissues has been elucidated by Renfret (1968) (Renfret A.P. Cryobiology: some fundamentals in surgical context. In: Cryosurgery. Rand R.W., Rinfret A.P., von Lode H., Eds. Springfield, IL: Charles C. Thomas, 1968) and by Mazur (1984): ice formation begins in the intercellular spaces. The vapor pressure of the ice is lower than the vapor pressure of the solute water in the surrounding cells and as heat is removed at the freezing point of the solutions, the ice crystals grow between the cells, extracting water from them. As the ice crystals grow, the volume of the cells shrinks, and the cells are crushed between the ice crystals. Additionally, as the cells shrink, the solutes inside the cells are concentrated in the remaining water, increasing the intracellular ionic strength and interfering with the organization of the proteins and other organized intercellular structures. Eventually, the solute concentration inside the cells reaches the eutectic and freezes. The final state of frozen tissues is pure ice in the former extracellular spaces, and inside the cell membranes a mixture of concentrated cellular components in ice and bound water. In general, this process is not reversible to the point of restoring the tissues to life.

Cryostasis utilizes using clathrate-forming gases that penetrate and saturate the biological tissues causing clathrate hydrates formation (under specific pressure-temperature conditions) inside the cells and in the extracellular matrix. Clathrate hydrates are a class of solids in which gas molecules occupy "cages" made up of hydrogen-bonded water molecules. These "cages" are unstable when empty, collapsing into conventional ice crystal structure, but they are stabilised by the inclusion of the gas molecule within them. Most low molecular weight gases (including CH_4, H_2S, Ar, Kr, and Xe) will form a hydrate under some pressure-temperature conditions. Clathrates formation will prevent the biological tissues from dehydration which will cause irreversible inactivation of intracellular enzymes.

Neuropreservation

Neuropreservation is a type of cryonics procedure where the brain is preserved with the intention of future resuscitation and regrowth of a healthy body around the brain. Usually the brain is left within the head for physical protection, so the whole head is cryopreserved. A cryonics patient who undergoes neuropreservation is said to be a neuropatient.

The procedure is often done because vitrification of the entire body is not yet available. Vitrification essentially eliminates the mechanical and chemical damage caused by ice formation, at the cost of cryoprotectant toxicity and side effects of dehydration of tissue due to the blood-brain barrier. Although not a direct consequence of vitrification itself, storage of the vitrified brain directly in liquid nitrogen raises the further aspect of fractures, which are fewer in number but larger in scale in vitrified tissue than frozen tissue, a consequence of cooling from T_g (-135 °C) to liquid nitrogen's boiling point (-196 °C).

Advantages

Neuropreservation has several advantages over whole body preservation. It costs less; neuropatients are easier to transport in case of legal, social, or physical problems; it is possible to do a better job of perfusing and therefore cryoprotecting the brain when there is no need to consider other tissues, and its smaller volume allows more rapid and less expensive cooling. Aubrey de Grey has theorized that neuropatients will be revived after procedures have been perfected on whole body patients, and therefore have better chances for revival.

History

Neuropreservation was first proposed in 1965 by cryonics co-creator Evan Cooper, proposed again in a speculative scientific paper by gerontologist George M. Martin in 1971, and independently proposed yet again in 1974 by Mike Darwin, and Fred and Linda Chamberlain. The Chamberlains were the founders of the Alcor Life Extension Foundation. In 1976 Fred's father became the first of many neuropreservation patients at Alcor.

Prior to the year 2000, neuropreservation was performed by surgical separation of the body from the head (called cephalic isolation or "neuroseparation") at the end of cryoprotectant perfusion performed on the upper body via the ascending aorta. After that year, Alcor began performing cephalic isolation before cryoprotectant perfusion, in deep hypothermia, and then using the carotid and vetebral arteries directly for perfusion with cryoprotectants.

As of 2014, Alcor, Oregon Cryonics, and KrioRus are the only cryonics organizations that offer neuropreservation. Other organizations, such as the other major provider, the Cryonics Institute, avoid it because they are concerned about neuro's negative effect on the public's perception of cryonics and, especially, because of the negative impact on the families of patients. Journalists & horror novelists invariably have a field day with "frozen severed heads," and focus not on the scientific or humanitarian purposes of cryonics, but on sensationalizing cryonics as grotesque or ridiculous. Their policy to always preserve the entire body prevents anyone leveling those claims at CI. Their goal is to preserve and eventually revive people, so they avoid processes like neurocryopreservation that could damage or otherwise put their patients at risk. Alcor claims there are good technical justifications for neuropreservation, and that they will continue to offer it. Approximately three quarters of the cryonics patients stored at Alcor are neuropatients.

References

- Tilden, William Augustus (2009). A Short History of the Progress of Scientific Chemistry in Our Own Times. BiblioBazaar, LLC. p. 249. ISBN 1-103-35842-1.
- Wainner, Scott; Richmond, Robert (2003). The Book of Overclocking: Tweak Your PC to Unleash Its Power. No Starch Press. p. 44. ISBN 1-886411-76-X.
- Moynihan, C. et al. in The Glass Transition and the Nature of the Glassy State, Eds. M. Goldstein and R. Simha, Ann. N.Y. Acad. Sci., Vol. 279 (1976) ISBN 0890720533
- John W. Nicholson (2011). The Chemistry of Polymers (4, Revised ed.). Royal Society of Chemistry. p. 50. ISBN 9781849733915. Retrieved 10 September 2013.
- Cowie, J. M. G. and Arrighi, V., Polymers: Chemistry and Physics of Modern Materials, 3rd Edn. (CRC Press, 2007) ISBN 0748740732
- Frankham, Dick; Ballou, Jon; Briscoe, David (2011). Introduction to Conservation Genetics. United Kingdom: Cabridge University Press. pp. 430–471. ISBN 978-0-521-70271-3.
- Yawn, David H. "Cryopreservation." Encyclopedia Britannica Online. Encyclopedia Britannica, n.d. Web. May 19, 2016. http://www.britannica.com/technology/cryopreservation
- Brockbank, Dr. Kelvin G. M., James C. Covault, and Dr. Michael J. Taylor. "Cryoconservation Guide." Dr. Michael J. Taylor. Web. May 16, 2016.
- "IUCN Species Survival Commission Guidelines on the Use of Ex situ Management for Species Conservation" (PDF). IUCN. 2014. Retrieved 27 May 2016.

3

Cryopreservation in Nature

Fishes and insects residing in the Arctic and Antarctic manufacture proteins as protection to minimize damage during the cold weather. This substance that is used to protect them is known as cryoprotectant. Freezing colds are tolerated by organisms through cold hardening procedures.

Antifreeze Protein

Antifreeze proteins (AFPs) or ice structuring proteins (ISPs) refer to a class of polypeptides produced by certain vertebrates, plants, fungi and bacteria that permit their survival in subzero environments. AFPs bind to small ice crystals to inhibit growth and recrystallization of ice that would otherwise be fatal. There is also increasing evidence that AFPs interact with mammalian cell membranes to protect them from cold damage. This work suggests the involvement of AFPs in cold acclimatization.

Non-colligative Properties

Unlike the widely used automotive antifreeze, ethylene glycol, AFPs do not lower freezing point in proportion to concentration. Rather, they work in a noncolligative manner. This phenomenon allows them to act as an antifreeze at concentrations 1/300th to 1/500th of those of other dissolved solutes. Their low concentration minimizes their effect on osmotic pressure. The unusual properties of AFPs are attributed to their selective affinity for specific crystalline ice forms and the resulting blockade of the ice-nucleation process.

Thermal Hysteresis

AFPs create a difference between the melting point and freezing point known as thermal hysteresis. The addition of AFPs at the interface between solid ice and liquid water inhibits the thermodynamically favored growth of the ice crystal. Ice growth is kinetically inhibited by the AFPs covering the water-accessible surfaces of ice.

Thermal hysteresis is easily measured in the lab with a nanolitre osmometer. Organisms differ in their values of thermal hysteresis. The maximum level of thermal hysteresis shown by fish AFP is approximately -1.5 °C (29.3 °F). However, insect antifreeze proteins are 10–30 times more active than fish proteins. This difference probably reflects the lower temperatures encountered by insects on land. In contrast, aquatic organisms are exposed only to -1 to -2 °C below freezing. During the extreme winter months, the spruce budworm resists freezing at temperatures approaching -30 °C. The Alaskan beetle *Upis ceramboides* can survive in a temperature of -60 °C by using antifreeze agents that are not proteins.

The rate of cooling can influence the thermal hysteresis value of AFPs. Rapid cooling can substantially decrease the nonequilibrium freezing point, and hence the thermal hysteresis value. Consequently, organisms cannot necessarily adapt to their subzero environment if the temperature drops abruptly.

Freeze Tolerance Versus Freeze Avoidance

Freeze avoidant: These species are able to prevent their body fluids from freezing altogether. Generally, the AFP function may be overcome at extremely cold temperatures, leading to rapid ice growth and death.

Freeze tolerant: These species are able to survive body fluid freezing. Some freeze tolerant species are thought to use AFPs as cryoprotectants to prevent the damage of freezing, but not freezing altogether. The exact mechanism is still unknown. However, it is thought AFPs may inhibit recrystallization and stabilize cell membranes to prevent damage by ice. They may work in conjunction with protein ice nucleators (PINs) to control the rate of ice propagation following freezing.

Diversity

There are many known nonhomologous types of AFPs.

Fish AFPs

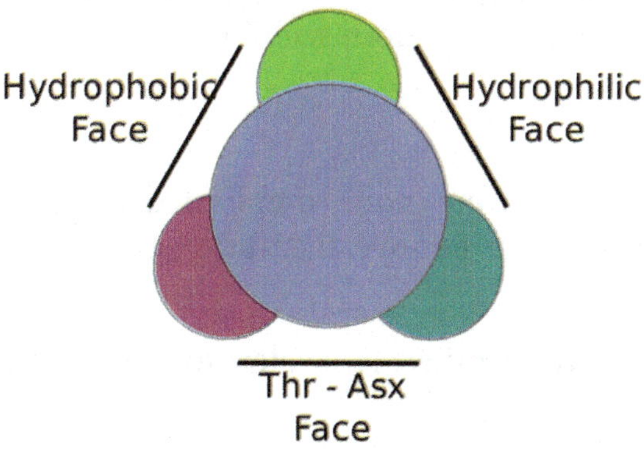

Figure 1. The three faces of Type I AFP

Antifreeze glycoproteins or AFGPs are found in Antarctic notothenioids and northern cod. They are 2.6-3.3 kD. AFGPs evolved separately in notothenioids and northern cod. In notothenioids, the AFGP gene arose from an ancestral trypsinogen-like serine protease gene.

Type I AFP is found in winter flounder, longhorn sculpin and shorthorn sculpin. It is the best documented AFP because it was the first to have its three-dimensional structure determined. Type I AFP consists of a single, long, amphipathic alpha helix, about 3.3-4.5 kD in size. There are three faces to the 3D structure: the hydrophobic, hydrophilic, and Thr-Asx face.

Type I-hyp AFP (where hyp stands for hyperactive) are found in several righteye flounders. It is approximately 32 kD (two 17 kD dimeric molecules). The protein was isolated from the blood plasma of winter flounder. It is considerably better at depressing freezing temperature than most fish AFPs.

Type II AFPs are found in sea raven, smelt and herring. They are cysteine-rich globular proteins containing five disulfide bonds. Type II AFPs likely evolved from calcium dependent (c-type) lectins. Sea ravens, smelt, and herring are quite divergent lineages of teleost. If the AFP gene were present in the most recent common ancestor of these lineages, it's peculiar that the gene is scattered throughout those lineages, present in some orders and absent in others. It has been suggested that lateral gene transfer could be attributed to this discrepancy, such that the smelt acquired the type II AFP gene from the herring.

Type III AFPs are found in Antarctic eelpout. They exhibit similar overall hydrophobicity at ice binding surfaces to type I AFPs. They are approximately 6kD in size. Type III AFPs likely evolved from a sialic acid synthase gene present in Antarctic eelpout. Through a gene duplication event, this gene—which has been shown to exhibit some ice-binding activity of its own—evolved into an effective AFP gene.

Type IV AFPs are found in longhorn sculpins. They are alpha helical proteins rich in glutamate and glutamine. This protein is approximately 12KDa in size and consists of a 4-helix bundle. Its only posttranslational modification is a pyroglutamate residue, a cyclized glutamine residue at its N-terminus. Scientists at the University of Guelph in Canada are currently examining the role of this pyroglutame residue in the antifreeze activity of type IV AFP from the longhorn sculpin.

Plant AFPs

The classification of AFPs became more complicated when antifreeze proteins from plants were discovered. Plant AFPs are rather different from the other AFPs in the following aspects:

1. They have much weaker thermal hysteresis activity when compared to other AFPs.
2. Their physiological function is likely in inhibiting the recrystallization of ice rather than in the preventing ice formation.
3. Most of them are evolved pathogenesis-related proteins, sometimes retaining antifungal properties.

Insect AFPs

There are two types of insect antifreeze proteins, *Tenebrio* and *Dendroides* AFPs which are both in different insect families. They are similar to one another, both being hyperactive (i.e. greater thermal hysteresis value) and consist of varying numbers of 12- or 13-mer repeats of approximately 8.3 to 12.5 kD. Throughout the length of the protein, at least every sixth residue is a cysteine.

Tenebrio or Type V AFPs are found in beetles, whereas *Dendroides* or *Choristoneura fumiferana* AFPs are found in some Lepidoptera.

Sea Ice Organisms Afps

AFPs were also found in microorganisms living in sea ice. The diatoms *Fragilariopsis cylindrus* and *F. curta* play a key role in polar sea ice communities, dominating the assemblages of both platelet layer and within pack ice. AFPs are widespread in these species, and the presence of AFP genes as a multigene family indicates the importance of this group for the genus *Fragilariopsis*. AFPs identified in *F. cylindrus* belong to an AFP family which is represented in different taxa and can be found in other organisms related to sea ice (*Colwellia* spp., *Navicula glaciei*, *Chaetoceros neogracile* and *Stephos longipes and Leucosporidium antarcticum*) and Antarctic inland ice bacteria (Flavobacteriaceae), as well as in cold-tolerant fungi (*Typhula ishikariensis*, *Lentinula edodes* and *Flammulina populicola*.)

Evolution

The remarkable diversity and distribution of AFPs suggest the different types evolved recently in response to sea level glaciation occurring 1-2 million years ago in the Northern hemisphere and 10-30 million years ago in Antarctica. This independent development of similar adaptations is referred to as convergent evolution. There are two reasons why many types of AFPs are able to carry out the same function despite their diversity:

1. Although ice is uniformly composed of oxygen and hydrogen, it has many different surfaces exposed for binding. Different types of AFPs may interact with different surfaces.

2. Although the five types of AFPs differ in their primary structure of amino acids, when each folds into a functioning protein, they may share similarities in their three-dimensional or tertiary structure that facilitates the same interactions with ice.

Mechanisms of Action

AFPs are thought to inhibit growth by an adsorption–inhibition mechanism. They adsorb to nonbasal planes of ice, inhibiting thermodynamically favored ice growth. The presence of a flat, rigid surface in some AFPs seems to facilitate its interaction with ice via Van der Waals force surface complementarity.

Binding to Ice

Normally, ice crystals grown in solution only exhibit the basal (0001) and prism faces (1010), and appear as round and flat discs. However, it appears the presence of AFPs exposes other faces. It now appears the ice surface 2021 is the preferred binding surface, at least for AFP type I. Through studies on type I AFP, ice and AFP were initially thought to interact through hydrogen bonding (Raymond and DeVries, 1977). However, when parts of the protein thought to facilitate this hydrogen bonding were mutated, the hypothesized decrease in antifreeze activity was not observed. Recent data suggest hydrophobic interactions could be the main contributor. It is difficult to discern the exact mechanism of binding because of the complex water-ice interface. Currently, attempts to uncover the precise mechanism are being made through use of molecular modelling programs (molecular dynamics or the Monte Carlo method).

Binding Mechanism and Antifreeze Function

According to the structure and function study on the antifreeze protein from the fish winter flounder, the antifreeze mechanism of the type-I AFP molecule was shown to be due to the binding to an ice nucleation structure in a zipper-like fashion through hydrogen bonding of the hydroxyl groups of its four Thr residues to the oxygens along the $[01\bar{1}2]$ direction in ice lattice, subsequently stopping or retarding the growth of ice pyramidal planes so as to depress the freeze point.

The above mechanism can be used to elucidate the structure-function relationship of other antifreeze proteins with the following two common features:

1. recurrence of a Thr residue (or any other polar amino acid residue whose side-chain can form a hydrogen bond with water) in an 11-amino-acid period along the sequence concerned, and
2. a high percentage of an Ala residue component therein.

History

In the 1950s, Norwegian scientist Scholander set out to explain how Arctic fish can survive in water colder than the freezing point of their blood. His experiments led him to believe there was “antifreeze” in the blood of Arctic fish. Then in the late 1960s, animal biologist Arthur DeVries was able to isolate the antifreeze protein through his investigation of Antarctic fish. These proteins were later called antifreeze glycoproteins (AFGPs) or antifreeze glycopeptides to distinguish them from newly discovered nonglycoprotein biological antifreeze agents (AFPs). DeVries worked with Robert Feeney

(1970) to characterize the chemical and physical properties of antifreeze proteins. In 1992, Griffith *et al.* documented their discovery of AFP in winter rye leaves. Around the same time, Urrutia, Duman and Knight (1992) documented thermal hysteresis protein in angiosperms. The next year, Duman and Olsen noted AFPs had also been discovered in over 23 species of angiosperms, including ones eaten by humans. As well, they reported their presence in fungi and bacteria.

Name Change

Recent attempts have been made to relabel antifreeze proteins as ice structuring proteins to more accurately represent their function and to dispose of any assumed negative relation between AFPs and automotive antifreeze, ethylene glycol. These two things are completely separate entities, and show loose similarity only in their function.

Commercial Applications

Numerous fields would be able to benefit from the protection of tissue damage by freezing. Businesses are currently investigating the use of these proteins in:

- Increasing freeze tolerance of crop plants and extending the harvest season in cooler climates
- Improving farm fish production in cooler climates
- Lengthening shelf life of frozen foods
- Improving cryosurgery
- Enhancing preservation of tissues for transplant or transfusion in medicine
- Therapy for hypothermia

Unilever has obtained UK approval to use a genetically modified yeast to produce antifreeze proteins from fish, for use in ice cream production.

Recent News

One recent, successful business endeavor has been the introduction of AFPs into ice cream and yogurt products. This ingredient, labelled ice-structuring protein, has been approved by the Food and Drug Administration. The proteins are isolated from fish and replicated, on a larger scale, in genetically modified yeast.

There is concern from organizations opposed to genetically modified organisms (GMOs), arguing modified antifreeze proteins may cause inflammation. Intake of non genetically modified AFPs in diet is likely substantial in most northerly and temperate regions already. Given the known historic consumption of AFPs, it is safe to conclude their functional properties do not impart any toxicologic or allergenic effects in humans.

As well, the transgenic process of ISP production is widely used in society already. Insulin and rennet are produced using this technology. The process does not impact the product; it merely makes production more efficient and prevents the death of fish which would otherwise be killed to extract the protein.

Currently, Unilever incorporates AFPs into some of its American products, including some popsicles and a new line of Breyers Light Double Churned ice cream bars. In ice cream, AFPs allow the production of very creamy, dense, reduced fat ice cream with fewer additives. They control ice crystal growth brought on by thawing on the loading dock or kitchen table which drastically reduces texture quality.

In November 2009, the Proceedings of the National Academy of Sciences published the discovery of a molecule in an Alaskan beetle that behaves like AFPs, but is composed of saccharides and fatty acids.

A 2010 study demonstrated the stability of superheated water ice crystals in an AFP solution, showing while the proteins can inhibit freezing, they can also inhibit melting.

Antifreeze

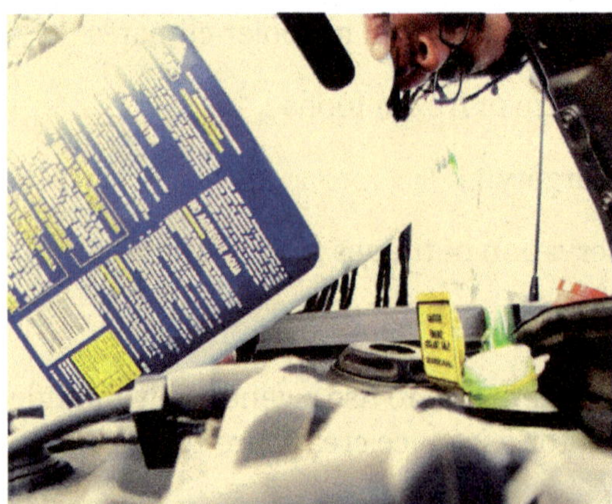

"Topping up" the antifreeze solution in a car's cooling system is a routine maintenance item for most modern cars.

An antifreeze is an additive which lowers the freezing point of a water-based liquid. An antifreeze mixture is used to achieve freezing-point depression for cold environments and also achieves boiling-point elevation ("anti-boil") to allow higher coolant temperature. Freezing and boiling points are colligative properties of a solution, which depend on the concentration of the dissolved substance.

Because water has good properties as a coolant, water plus antifreeze is used in internal combustion engines and other heat transfer applications, such as HVAC chillers and solar water heaters. The purpose of antifreeze is to prevent a rigid enclosure from bursting due to expansion when water freezes. Commercially, both the *additive* (pure concentrate) and the *mixture* (diluted solution) are called antifreeze, depending on the context. Careful selection of an antifreeze can enable a wide temperature range in which the mixture remains in the liquid phase, which is critical to efficient heat transfer and the proper functioning of heat exchangers.

Salts are frequently used for de-icing, but salt solutions are not used for cooling systems because they can cause severe corrosion to metals. Instead, non-corrosive antifreezes are commonly used for critical de-icing, such as for aircraft wings.

Automotive and Internal Combustion Engine Use

Fluorescent green-dyed antifreeze is visible in the radiator header tank when car radiator cap is removed

Most automotive engines are "water"-cooled to remove waste heat, although the "water" is actually antifreeze/water mixture and not plain water. The term engine coolant is widely used in the automotive industry, which covers its primary function of convective heat transfer for internal combustion engines. When used in an automotive context, corrosion inhibitors are added to help protect vehicles' radiators, which often contain a range of electrochemically incompatible metals (aluminum, cast iron, copper, brass, solder, et cetera). Water pump seal lubricant is also added.

Antifreeze was developed to overcome the shortcomings of water as a heat transfer fluid. In some engines freeze plugs (engine block expansion plugs) are placed in areas of the engine block where coolant flows in order to protect the engine from freeze damage if the ambient temperature drops below the freezing point of the antifreeze/water mixture. These should not be confused with core plugs, whose purpose is to allow removal of sand used in the casting process of engine blocks (core plugs will be pushed out if the coolant freezes, though, assuming that they adjoin the coolant passages, which is not always the case).

On the other hand, if the engine coolant gets too hot, it might boil while inside the engine, causing voids (pockets of steam), leading to localized hot spots and the catastrophic failure of the engine. If plain water were to be used as an engine coolant, it would promote galvanic corrosion. Proper engine coolant and a pressurized coolant system can help obviate the problems which make plain water incompatible with automotive engines. With proper antifreeze, a wide temperature range can be tolerated by the engine coolant, such as −34 °F (−37 °C) to +265 °F (129 °C) for 50% (by volume) propylene glycol diluted with water and a 15 psi pressurized coolant system.

Early engine coolant antifreeze was methanol (methyl alcohol). Methanol was widely used in windshield fluids, however, in Europe, due to new REACH legislation, the use of methanol in windshield fluids is limited to 5% and in the near future will be further reduced to 3%. As radiator caps were vented, not sealed, the methanol was lost to evaporation, requiring frequent replenishment to avoid freezing of the coolant. Methanol also accelerates corrosion of the metals, especially aluminum, used in the engine and cooling systems. Ethylene glycol was developed, and soon replaced methanol as an engine cooling system antifreeze. It has a very low volatility compared to methanol and to water. Before the 1950s, coolant systems were unpressurized and the engine was often cooler than modern automotive engines. By pressurizing the coolant system with a radiator cap, the boiling point of the fluid is increased, permitting higher engine temperatures and better fuel efficiency. Pressurized systems do not appreciably change the freezing point.

Other Uses

The most common water-based antifreeze solutions used in electronics cooling are mixtures of water and either ethylene glycol (EGW) or propylene glycol (PGW). The use of ethylene glycol has a longer history, especially in the automotive industry. However, EGW solutions formulated for the automotive industry often have silicate based rust inhibitors that can coat and/or clog heat exchanger surfaces. Ethylene glycol is listed as a toxic chemical requiring care in handling and disposal.

Ethylene glycol has desirable thermal properties, including a high boiling point, low freezing point, stability over a wide range of temperatures, and high specific heat and thermal conductivity. It also has a low viscosity and, therefore, reduced pumping requirements. Although EGW has more desirable physical properties than PGW, the latter coolant is used in applications where toxicity might be a concern. PGW is generally recognized as safe for use in food or food processing applications, and can also be used in enclosed spaces.

Similar mixtures are commonly used in HVAC and industrial heating or cooling systems as a high-capacity heat transfer medium. Many formulations have corrosion inhibitors, and it is expected that these chemicals will be replenished (manually or under automatic control) to keep expensive piping and equipment from corroding.

Primary Agents

Most antifreeze is made by mixing distilled water with some kind of alcohol.

Methanol

Methanol (also known as methyl alcohol, carbinol, wood alcohol, wood naphtha or wood spirits) is a chemical compound with chemical formula CH_3OH. It is the simplest alcohol, and is a light, volatile, colorless, flammable, poisonous liquid with a distinctive odor that is somewhat milder and sweeter than ethanol (ethyl alcohol). At room temperature, it is a polar solvent and is used as an antifreeze, solvent, fuel, and as a denaturant for ethyl alcohol. It is not popular for machinery, but may be found in automotive windshield washer fluid, de-icers, and gasoline additives.

Ethylene Glycol

HO OH

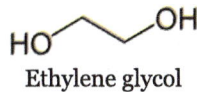

Ethylene glycol

Ethylene glycol solutions became available in 1926 and were marketed as "permanent antifreeze" since the higher boiling points provided advantages for summertime use as well as during cold weather. They are used today for a variety of applications, including automobiles, but gradually being replaced by propylene glycol due to its lower toxicity.

When ethylene glycol is used in a system, it may become oxidized to five organic acids (formic, oxalic, glycolic, glyoxalic and acetic acid). Inhibited ethylene glycol antifreeze mixes are available, with additives that buffer the pH and reserve alkalinity of the solution to prevent oxidation of ethylene glycol and formation of these acids. Nitrites, silicates, theodin, borates and azoles may also be used to prevent corrosive attack on metal.

Poisoning

Ethylene glycol is poisonous to humans and other animals, and should be handled carefully and disposed of properly. Its sweet taste can lead to accidental ingestion or allow its deliberate use as a murder weapon. Ethylene glycol is difficult to detect in the body, and causes symptoms—including intoxication, severe diarrhea, and vomiting—that can be confused with other illnesses or diseases. Its metabolism produces calcium oxalate, which crystallizes in the brain, heart, lungs, and kidneys, damaging them; depending on the level of exposure, accumulation of the poison in the body can last weeks or months before causing death, but death by acute kidney failure can result within 72 hours if the individual does not receive appropriate medical treatment for the poisoning. Some ethylene glycol antifreeze mixtures contain an embittering agent, such as denatonium, to discourage accidental or deliberate consumption.

Propylene Glycol

Propylene glycol, on the other hand, is considerably less toxic than ethylene glycol and may be labeled as "non-toxic antifreeze". It is used as antifreeze where ethylene glycol would be inappropriate, such as in food-processing systems or in water pipes in homes where incidental ingestion may be possible. As confirmation of its relative non-toxicity, the FDA allows propylene glycol to be added to a large number of processed foods, including ice cream, frozen custard, salad dressings and baked goods.

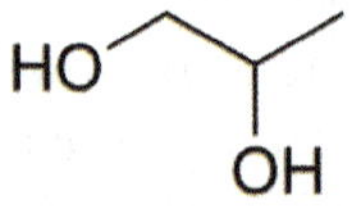

HO OH

Propylene glycol

Propylene glycol oxidizes when exposed to air and heat, forming lactic acid. If not properly inhibited, this fluid can be very corrosive, so pH buffering agents such as dipotassium phosphate, Protodin and potassium bicarbonate are often added to propylene glycol, to prevent acidic corrosion of metal components. Pre-inhibited propylene glycol solutions like Dowfrost (manufactured by Dow Chemicals, US) and Tonofrost (manufactured by Chemtex Speciality Ltd, India) can also be used instead of pure propylene glycol to prevent corrosion.

Besides cooling system corrosion, biological fouling also occurs. Once bacterial slime starts to grow, the corrosion rate of the system increases. Maintenance of systems using glycol solution includes regular monitoring of freeze protection, pH, specific gravity, inhibitor level, color, and biological contamination. Propylene glycol should be replaced when it turns a reddish color.

Glycerol

Once used for automotive antifreeze, glycerol has the advantage of being non-toxic, withstands relatively high temperatures, and is noncorrosive.

Like ethylene glycol and propylene glycol, glycerol is a non-ionic kosmotrope that forms strong hydrogen bonds with water molecules, competing with water-water hydrogen bonds. This disrupts the crystal lattice formation of ice unless the temperature is significantly lowered. The minimum freezing point temperature is at about −36 °F / −37.8 °C corresponding to 60–70% glycerol in water.

Glycerol was historically used as an antifreeze for automotive applications before being replaced by ethylene glycol, which has a lower freezing point. While the minimum freezing point of a glycerol-water mixture is higher than an ethylene glycol-water mixture, glycerol is not toxic and is being re-examined for use in automotive applications. Glycerol is mandated for use as an antifreeze in many sprinkler systems.

In the laboratory, glycerol is a common component of solvents for enzymatic reagents stored at temperatures below 0 °C due to the depression of the freezing temperature of solutions with high concentrations of glycerol. It is also used as a cryoprotectant where the glycerol is dissolved in water to reduce damage by ice crystals to laboratory organisms that are stored in frozen solutions, such as bacteria, nematodes, and mammalian embryos.

Measuring the Freeze Point

Once antifreeze has been mixed with water and put into use, it periodically needs to be maintained. If engine coolant leaks, boils, or if the cooling system needs to be drained and refilled, the antifreeze's freeze protection will need to be considered. In other cases a vehicle may need to be operated in a colder environment, requiring more antifreeze and less water. Three methods are commonly employed to determine the freeze point of the solution:

1. Specific gravity—(using a hydrometer or some sort of floating indicator),
2. Refractometer—which measures the refractive index of the antifreeze solution and translates it into freeze point, and
3. Test strips—specialized, disposable indicators made for this purpose.

Although ethylene glycol hydrometers are widely available and mass-marketed for antifreeze testing, they give false readings at high temperatures because specific gravity changes with temperature. Propylene glycol solutions cannot be tested using specific gravity because of ambiguous results (40% and 100% solutions have the same specific gravity).

Corrosion Inhibitors

Most commercial antifreeze formulations include corrosion inhibiting compounds, and a colored dye (commonly a fluorescent green, red, orange, yellow, or blue) to aid in identification. A 1:1 dilution with water is usually used, resulting in a freezing point of about −34 °F (−37 °C), depending on the formulation. In warmer or colder areas, weaker or stronger dilutions are used, respectively, but a range of 40%/60% to 60%/40% is frequently specified to ensure corrosion protection, and 70%/30% for maximum freeze prevention down to −84 °F (−64 °C).

Maintenance

In the absence of leaks, antifreeze chemicals such as ethylene glycol or propylene glycol may retain their basic properties indefinitely. By contrast, corrosion inhibitors are gradually used up, and must be replenished from time to time. Larger systems (such as HVAC systems) are often monitored by specialist firms which take responsibility for adding corrosion inhibitors and regulating coolant composition. For simplicity, most automotive manufacturers recommend periodic complete replacement of engine coolant, to simultaneously renew corrosion inhibitors and remove accumulated contaminants.

Traditional Inhibitors

Traditionally, there were two major corrosion inhibitors used in vehicles: silicates and phosphates. American made vehicles traditionally used both silicates and phosphates.

European makes contain silicates and other inhibitors, but no phosphates. Japanese makes traditionally use phosphates and other inhibitors, but no silicates.

Organic Acid Technology

Certain cars are built with organic acid technology (OAT) antifreeze (e.g., DEX-COOL), or with a hybrid organic acid technology (HOAT) formulation (e.g., Zerex G-05), both of which are claimed to have an extended service life of five years or 240,000 km (150,000 mi).

DEX-COOL specifically has caused controversy. Litigation has linked it with intake manifold gasket failures in General Motors' (GM's) 3.1L and 3.4L engines, and with other failures in 3.8L and 4.3L engines. One of the anti-corrosion components presented as sodium or Potassium 2-ethylhexanoate and ethylhexanoic acid is incompatible with nylon 6,6 and silicone rubber, and is a known plasticizer. Class action lawsuits were registered in several states, and in Canada, to address some of these claims. The first of these to reach a decision was in Missouri, where a settlement was announced early in December 2007. Late in March 2008, GM agreed to compensate complainants in the remaining 49 states. GM (Motors Liquidation Company) filed for bankruptcy in 2009, which tied up the outstanding claims until a court determines who gets paid.

According to the DEX-COOL manufacturer, "mixing a 'green' [non-OAT] coolant with DEX-COOL reduces the batch's change interval to 2 years or 30,000 miles, but will otherwise cause no damage to the engine". DEX-COOL antifreeze uses two inhibitors: sebacate and 2-EHA (2-ethylhexanoic acid), the latter which works well with the hard water found in the United States, but is a plasticizer that can cause gaskets to leak.

According to internal GM documents, the ultimate culprit appears to be operating vehicles for long periods of time with low coolant levels. The low coolant is caused by pressure caps that fail in the open position. (The new caps and recovery bottles were introduced at the same time as DEX-COOL). This exposes hot engine components to air and vapors, causing corrosion and contamination of the coolant with iron oxide particles, which in turn can aggravate the pressure cap problem as contamination holds the caps open permanently.

Honda and Toyota's new extended life coolant use OAT with sebacate, but without the 2-EHA. Some added phosphates provide protection while the OAT builds up. Honda specifically excludes 2-EHA from their formulas.

Typically, OAT antifreeze contains an orange dye to differentiate it from the conventional glycol-based coolants (green or yellow). Some of the newer OAT coolants claim to be compatible with *all* types of OAT and glycol-based coolants; these are typically green or yellow in color.

Hybrid Organic Acid Technology

HOAT coolants typically mix an OAT with a traditional inhibitor, such as silicates or phosphates.

G05 is a low-silicate, phosphate free formula that includes the benzoate inhibitor.

Additives

All automotive antifreeze formulations, including the newer organic acid (OAT antifreeze) formulations, are environmentally hazardous because of the blend of additives (around 5%), including lubricants, buffers and corrosion inhibitors. Because the additives in antifreeze are proprietary, the safety data sheets (SDS) provided by the manufacturer list only those compounds which are considered to be significant safety hazards when used in accordance with the manufacturer's recommendations. Common additives include sodium silicate, disodium phosphate, sodium molybdate, sodium borate, denatonium benzoate and dextrin (hydroxyethyl starch). Disodium fluorescein dyes are added to antifreeze to help trace the source of leaks, and as an identifier since some different formulations are incompatible.

Automotive antifreeze has a characteristic odor due to the additive tolytriazole, a corrosion inhibitor. The unpleasant odor in industrial use tolytriazole comes from impurities in the product that are formed from the toluidine isomers (ortho-, meta- and para-toluidine) and meta-diamino toluene which are side-products in the manufacture of tolytriazole. These side-products are highly reactive and produce volatile aromatic amines which are responsible for the unpleasant odor.

Psychrophile

Psychrophiles or *cryophiles* (adj. cryophilic) are extremophilic organisms that are capable of growth and reproduction in cold temperatures, ranging from −20 °C to +10 °C. Temperatures as low as −15 °C are found in pockets of very salty water (brine) surrounded by sea ice. Psychrophiles are true extremophiles because they adapt not only to low temperatures but often also to further environmental constraints. They can be contrasted with thermophiles, which thrive at unusually hot temperatures. In addition to that, distinctions between mesophilic and psychrophilic cold-shock response, including lack of repression of house-keeping protein synthesis and the presence of cold-acclimation proteins (Caps) in psychrophiles, does exist. The environments they inhabit are ubiquitous on Earth, as a large fraction of our planetary surface experiences temperatures lower than 15 °C. They are present in alpine and arctic soils, high-latitude and deep ocean waters, polar ice, glaciers, and snowfields. They are of particular interest to astrobiology, the field dedicated to the formulation of theory about the possibility of extraterrestrial life, and to geomicrobiology, the study of microbes active in geochemical processes. In experimental work at University of Alaska Fairbanks, a 1000-litre biogas digester using psychrophiles harvested from "mud from a frozen lake in Alaska" has produced 200–300 litres of methane per day, about 20–30% of the output from digesters in warmer climates.

Psychrophiles use a wide variety of metabolic pathways, including photosynthesis, chemoautotrophy (also sometimes known as lithotrophy), and heterotrophy, and form robust, diverse communities. Most psychrophiles are bacteria or archaea, and psychrophily is present in widely diverse microbial lineages within those broad groups. Additionally, recent research has discovered novel groups of psychrophilic fungi living in oxygen-poor areas under alpine snowfields. A further group of eukaryotic cold-adapted organisms are snow algae, which can cause watermelon snow. Some multicellular eukaryotes can also be metabollicaly active at negative temperatures, such as some conifers that can still photosynthetize when it is several degrees under 0 °C (conifers are known to be often more cold-resistant than broadleaf trees). Psychrophiles are interesting enzymes that are very useful models in the research of proteins. Psychrophiles are characterized by lipid cell membranes chemically resistant to the stiffening caused by extreme cold, and often create protein 'antifreezes' to keep their internal space liquid and protect their DNA even in temperatures below water's freezing point. A commonly accepted hypothesis for this cold adaptation is the activity-stability-flexibility relationship, suggesting that psychrophilic enzymes increase the flexibility of their structure to compensate for the 'freezing effect' of cold habitats.

Examples are *Arthrobacter* sp., *Psychrobacter* sp. and members of the genera *Halomonas*, *Pseudomonas*, *Hyphomonas*, and *Sphingomonas*. Another example is the *Chryseobacterium greenlandensis*, a psychrophile that was found in 120,000 years old ice.

Psychrophiles vs. Psychrotrophs

In 1940, ZoBell and Conn stated that they have never encountered "true psychrophiles" or organisms that grow best at relatively low temperatures. In 1958, J. L. Ingraham supported this by concluding that there are very few or possibly no bacteria that fit the textbook definitions of psychrophiles. Richard Y. Morita emphasizes this by using the term "psychrotrophic" to describe organisms that do not meet the definition of psychrophiles. The confusion between the terms Psychrotrophs and psychrophiles was started because investigators were unaware of the thermolability of psychrophilic organisms at the laboratory temperatures. Due to this, early investigators did not determine the cardinal temperatures for their isolates. The similarity between these two is that they are both capable of growing at zero, but optimum and upper temperature limits for the growth are lower for psychrophiles compared to psychrotrophs. Psychrophiles are also more often isolated from permanently cold habitats compared to psychrotrophs. Although psychrophilic enzymes remain under-used because the cost of production and processing at low temperatures is higher than for the commercial enzymes that are presently in use, the attention and resurgence of research interest in psychrophiles and psychrotrophs will be a contributor to the betterment of the environment and the desire to conserve energy.

Insect Winter Ecology

Insect winter ecology entails the overwinter survival strategies of insects, which are in many respects more similar to those of plants than to many other animals, such as mammals and birds. This is because unlike those animals, which can generate their own heat internally (endothermic), insects must rely on external sources to provide their heat (ectothermic). Thus, insects sticking around in the winter, must tolerate freezing or rely on other mechanisms to avoid freezing. Loss of enzymatic function and eventual freezing due to low temperatures daily threatens the livelihood of these organisms during winter. Not surprisingly, insects have evolved a number of strategies to deal with the rigors of winter temperatures in places where they would otherwise not survive.

Survival Strategies

Two major strategies for winter survival have evolved in the Class Insecta due to their inability to generate significant heat metabolically. The first, migration, is a complete avoidance of the temperatures that pose a threat. If an insect cannot migrate, then it must stay and deal with the cold temperatures in one of two ways. This cold hardiness is separated into two categories, freeze avoidance and freeze tolerance.

Migration

Migration in insects is different than in birds. Bird migration is a two-way, round-trip movement of each individual, whereas this is not usually the case with insects. The short lifespan of insects compared to birds means that the adult that made one leg of the trip will be replaced by a member of the next generation on the return voyage. As a result, invertebrate biologists have redefined migration for this group of organisms as consisting of three parts:

1. A persistent, straight line movement away from the natal area
2. Distinctive pre- and post-movement behaviors
3. Re-allocation of energy within the body associated with the movement

This definition allows for mass insect movements to be considered as migration. Perhaps the best known insect migration is that of the monarch butterfly. The monarch in

North America migrates from as far north as Canada southward to Mexico and Southern California annually from about August to October. The population east of the Rocky Mountains overwinters in Michoacán, Mexico, and the western population overwinters in various sites in central coastal California, notably in Pacific Grove and Santa Cruz. The round trip journey is typically around 3,600 km in length. The longest one-way flight on record for monarchs is 3,009 km from Ontario, Canada to San Luis Potosí, Mexico. They use the direction of sunlight and magnetic cues to orient themselves during migration.

The monarch requires significant energy to make such a long flight, which is provided by fat reserves. When they reach their overwintering sites, they begin a period of lowered metabolic rate. Nectar from flowers procured at the overwintering site provides energy for the northward migration. To limit their energy use, monarchs congregate in large clusters in order to maintain a suitable temperature. This strategy, similar to huddling in small mammals, makes use of body heat from all the organisms and lowers heat loss.

Another common winter migrant insect, found in much of North America, South America, and the Caribbean, is the Green Darner. Migration patterns in this species are much less studied than those of monarchs. Green darners leave their northern ranges in September and migrate south. Studies have noted a seasonal influx of green darners to southern Florida, which indicates migratory behavior. Little has been done with tracking of the green darner, and reasons for migration are not fully understood since there are both resident and migrant populations. The common cue for migration southward in this species is the onset of winter.

Freeze Avoidance

Lethal freezing occurs when insects are exposed to temperatures below the melting point (MP) of their body fluids; therefore, insects that do not migrate from regions with the onset of colder temperatures must devise strategies to either tolerate or avoid freezing of intracellular and extracellular body fluids. Surviving colder temperatures, in insects, generally falls under two categories: Freeze-tolerant insects can tolerate the formation of internal ice and freeze-avoidant insects avoid freezing by keeping the bodily fluids liquid. The general strategy adopted by insects also differs between the northern hemisphere and the southern hemisphere. In temperate regions of the northern hemisphere where cold temperatures are expected seasonally and are usually for long periods of time, the main strategy is freeze avoidance. In temperate regions of the southern hemisphere, where seasonal cold temperatures are not as extreme or long lasting, the main strategy is freeze tolerance. However, in the Arctic, where freezing occurs seasonally, and for extended periods (>9 months), freeze tolerance also predominates.

Freeze avoidance involves both physiological and biochemical mechanisms. One method of freeze avoidance is the selection of a dry hibernation site in which no ice nucleation from an external source can occur. Insects may also have a physical barrier such as a wax-coated cuticle that provides protection against external ice across the cuticle. The stage of development at which an insect over-winters varies across species, but can occur at any point of the life cycle (i.e., egg, pupa, larva, and adult).

Freeze-avoidant insects that cannot tolerate the formation of ice within their bodily fluids need to implement strategies to depress the temperature at which their bodily fluids will freeze. Supercooling is the process by which water cools below its freezing point without changing phase into a solid, due to the lack of a nucleation source. Water requires a particle such as dust in order to crystallize and if no source of nucleation is introduced, water can cool down to -42°C without freezing. In the initial phase of seasonal cold hardening, ice-nucleating agents (INAs) such as food particles, dust particles and bacteria, in the gut or intracellular compartments of freeze avoidant insects have to be removed or inactivated. Removal of ice-nucleating material from the gut can be achieved by cessation in feeding, clearing the gut and removing lipoprotein ice nucleators (LPINs) from the haemolymph.

Overwintering lesser stag beetle larva

And in some species, by the shedding of the mid-gut during moulting.

In addition to physical preparations for winter, many insects also alter their biochemistry and metabolism. For example, some insects synthesize cryoprotectants such as polyols and sugars, which reduce the lethal freezing temperature of the body. Although polyols such as sorbitol, mannitol, and ethylene glycol can also be found, glycerol is by far the most common cryoprotectant and can be equivalent to ~20% of the total body mass. Glycerol is distributed uniformly throughout the head, the thorax, and the abdomen of insects, and is in equal concentration in intracellular and extracellular compartments. The depressive effect of glycerol on the super cooling point (SCP) is thought to be due to the high viscosity of glycerol solutions at low temperatures. This would inhibit INA activity and SCPs would drop far below the environmental temperature. At colder temperatures (below 0 °C), glycogen production is inhibited, and the breakdown of glycogen into glycerol is enhanced, resulting in the glycerol levels in freeze avoidant insects reaching levels five times higher than those in freeze tolerant insects which do not need to cope with extended periods of cold temperatures.

Though not all freeze avoidant insects produce polyols, all hibernating insects produce thermal hysteresis factors (THFs). A seasonal photoperiodic timing mechanism is responsible for increasing the antifreeze protein levels with concentrations reaching their highest in the winter. In the pyrochroid beetle, "Dendroides canadensis", a short photoperiod of 8 hours light and 16 hours of darkness, results in the highest levels of THFs, which corresponds with the shortening of daylight hours associated with winter. These antifreeze proteins are thought to stabilize SCPs by binding directly to the surface structures of the ice crystals themselves, diminishing crystal size and growth. Therefore, instead of acting to change the biochemistry of the bodily fluids as seen with cryoprotectants, THFs act directly with the ice crystals by adsorbing to the developing crystals to inhibit their growth and reduce the chance of lethal freezing occurring.

Freeze Tolerance

Freeze tolerance in insects refers to the ability of some insect species to survive ice formation within their tissues. All insects are ectothermic, which can make them vulnerable to freezing. In most animals, intra- and extracellular freezing causes severe tissue damage, resulting in death. Insects that have evolved freeze-tolerance strategies manage to avoid tissue damage by controlling where, when, and to what extent ice forms. In contrast to freeze avoiding insects that are able to exist in cold conditions by supercooling, freeze tolerant organisms limit supercooling and initiate the freezing of their body fluids at relatively high temperatures. Physiologically, this is accomplished through inoculative freezing, the production of ice nucleating proteins, crystalloid compounds, and/or microbes.

Although freeze-avoidance strategies predominate in the insects, freeze tolerance has evolved at least six times within this group (in the Lepidoptera, Blattodea, Diptera, Orthoptera, Coleoptera, and Hymenoptera). Freeze tolerance is also more prevalent in insects from the Southern Hemisphere (reported in 85% of species studied) than it is in insects from the Northern Hemisphere (reported in 29% of species studied). It has been suggested that this may be due to the Southern Hemisphere's greater climate variability, where insects must be able to survive sudden cold snaps yet take advantage of unseasonably warm weather as well. This is in contrast to the Northern Hemisphere, where predictable weather makes it more advantageous to overwinter after extensive seasonal cold hardening.

Examples of freeze tolerant insects include: the woolly bear, *Pyrrharctia isabella;* the flightless midge, *Belgica antarctica;* and the alpine cockroach, *Celatoblatta quinquemaculata.*

Dangers of Freezing

With some exceptions, the formation of ice within cells generally causes cell death even in freeze-tolerant species due to physical stresses exerted as ice crystals expand. Ice formation in extracellular spaces is also problematic, as it removes water from solution through the process of osmosis, causing the cellular environment to become hypertonic and draw water from the cell interiors. Excessive cell shrinkage can cause severe damage. This is because as ice forms outside the cell, the possible shapes that can be assumed by the cells are increasingly limited, causing damaging deformation. Finally, the expansion of ice within vessels and other spaces can cause physical damage to structures and tissues.

Ice Nucleators

In order for a body of water to freeze, a nucleus must be present upon which an ice crystal can begin to grow. At low temperatures, nuclei may arise spontaneously from clusters of slow-moving water molecules. Alternatively, substances that facilitate the aggregation of water molecules can increase the probability that they will reach the critical size necessary for ice formation.

Freeze-tolerant insects are known to produce ice nucleating proteins. The regulated production of ice nucleating proteins allows insects to control the formation of ice crystals within their bodies. The lower an insects' body temperature, the more likely it is that ice will begin to form spontaneously. Even freeze-tolerant animals cannot tolerate a sudden, total freeze; for most freeze-tolerant insects it is important that they avoid supercooling and initiate ice formation at relatively warm temperatures. This allows the insect to moderate the rate of ice growth, adjust more slowly to the mechanical and osmotic pressures imposed by ice formation.

Nucleating proteins may be produced by the insect, or by microorganisms that have become associated with the insects' tissues. These microorganisms possess proteins within their cell walls that function as nuclei for ice growth.

The temperature that a particular ice nucleator initiates freezing varies from molecule to molecule. Although an organism may possess a number of different ice nucleating proteins, only those that initiate freezing at the highest temperature will catalyze an ice nucleation event. Once freezing is initiated, ice will spread throughout the insect's body.

Cryoprotectants

The formation of ice in the extracellular fluid causes an overall movement of water out of cells, a phenomenon known as osmosis. As too much dehydration can be dangerous to cells, many insects possess high concentrations of solutes such as glycerol. Glycerol is a relatively polar molecule and therefore attracts water molecules, shifting the osmotic balance and holding some water inside the cells. As a result, cryoprotectants like glycerol decrease the amount of ice that forms outside of cells and reduce cellular dehydration. Insect cryoprotectants are also important for species that avoid freezing;

Intracellular Freezing

Most freeze-tolerant species restrict ice formation to extracellular spaces. Some species, however, can tolerate intracellular freezing as well. This was first discovered in the fat body cells of the goldenrod gall fly *Eurosta solidaginis*. The fat body is an insect tissue that is important for lipid, protein and carbohydrate metabolism (analogous to the mammalian liver). Although it is not certain why intracellular freezing is restricted to the fat body tissue in some insects, there is evidence that it may be due to the low water content within fat body cells.

Locations of Hibernating Insects

Insects are well hidden in winter, but there are several locations in which they can reliably be found. Ladybugs practice communal hibernation by stacking one on top of one another on stumps and under rocks to share heat and buffer themselves against winter temperatures. The female grasshopper (family Tettigoniidae [long-horned]), in an attempt to keep her eggs safe through the winter, tunnels into the soil and deposits her eggs as deep as possible in the ground. Many other insects, including various butterflies and moths also overwinter in soil in the egg stage. Some adult beetles hibernate underground during winter; many flies overwinter in the soil as pupae. Other methods of hibernation include the inhabitance of bark, where insects nest more toward the southern side of the tree for heat provided by the sun. Cocoons, galls, and parasitism are also common methods of hibernation.

Aquatic Insects

Insects that live under the water have different strategies for dealing with freezing than do terrestrial insects. Many insect species survive winter not as adults on land, but as larvae underneath the surface of the water. Under the water many benthic invertebrates will experience some subfreezing temperatures, especially in small streams. Aquatic insects have developed freeze tolerance much like their terrestrial counterparts. However, freeze avoidance is not an option for aquatic insects as the presence of ice in their surroundings may cause ice nucleation in their tissues. Aquatic insects have supercooling points typically around – 3° to – 7°C. In addition to using freeze tolerance, many aquatic insects migrate deeper into the water body where the temperatures are higher than at the surface. Insects such as stoneflies, mayflies, caddisflies, and dragonflies are common overwintering aquatic insects. The dance fly larvae have the lowest reported supercooling point for an aquatic insect at – 22°C.

References

- Danks, H.V. (1981). Arctic Arthropods. Ottawa, Canada: Entomological Society of Canada. p. 279. ISBN 0-9690829-0-8.
- Hudgens, R. Douglas; Hercamp, Richard D.; Francis, Jaime; Nyman, Dan A.; Bartoli, Yolanda

(2007). "An Evaluation of Glycerin (Glycerol) as a Heavy Duty Engine Antifreeze/Coolant Base". doi:10.4271/2007-01-4000. Retrieved 2013-06-07.

- A safe and effective propylene glycol based capture liquid for fruit fly traps baited with synthetic lures – page 2|Florida Entomologist. Findarticles.com. Retrieved on 2011-01-01.
- Gupta, Sujata (2010-11-06). "Biogas comes in from the cold". New Scientist. London: Sunita Harrington. p. 14. Retrieved 2011-02-04.

4

Cryogenics: An Overview

Cryogenics is the study of materials and their characteristics at very low temperatures. Refrigeration is a very common example of a cryogenic procedure. Cryogenic treatments enhance the quality of metals and other material. This chapter lists all the main applications of cryogenics and the processes that are involved.

Cryogenics

In physics, cryogenics is the study of the production and behaviour of materials at very low temperatures.

It is not well-defined at what point on the temperature scale refrigeration ends and cryogenics begins, but scientists assume it starts at or below −150 °C (123 K; −238 °F). The U.S. National Institute of Standards and Technology has chosen to consider the field of cryogenics as that involving temperatures below −180 °C or −292.00 °F or 93.15 K. This is a logical dividing line, since the normal boiling points of the so-called permanent gases (such as helium, hydrogen, neon, nitrogen, oxygen, and normal air) lie below −180 °C while the Freon refrigerants, hydrogen sulfide, and other common refrigerants have boiling points above −180 °C. (above −150 °C, −238 °F or 123 K).

A person who studies elements that have been subjected to extremely cold temperatures is called a cryogenicist.

Cryogenicists use the Kelvin or Rankine temperature scales.

Definitions and Distinctions

Cryogenics

> The branches of physics and engineering that involve the study of very low temperatures, how to produce them, and how materials behave at those temperatures.

Cryobiology

> The branch of biology involving the study of the effects of low temperatures on organisms (most often for the purpose of achieving cryopreservation).

Cryoconservation of animal genetic resources

The conservation of genetic material with the intention of conserving a breed.

Cryosurgery

The branch of surgery applying very low temperatures (down to −196 °C) to destroy malignant tissue, e.g. cancer cells.

Cryoelectronics

The field of research regarding superconductivity at low temperatures.

Cryotronics

The practical application of cryoelectronics.

Cryonics

Cryopreserving humans and animals with the intention of future revival. "Cryogenics" is sometimes erroneously used to mean "Cryonics" in popular culture and the press.

Etymology

The word *cryogenics* stems from Greek *cryo* – "cold" + *genic* – "having to do with production".

Cryogenic Fluids

Cryogenic fluids with their boiling point in kelvin (Liquid Helium3- 3.19k),(Liquid Helium4- 4.214k),(Liquid Hydrogen- 20.27k),(Liquid Neon-27.09k),(Liquid Nitrogen-77.36k),(Liquid Air- 78.8k),(Liquid Fluorine- 85.24k),(Liquid Argon- 87.24k),(Liquid Oxygen- 90.18),(Liquid Methane- 111.7k)

Industrial Applications

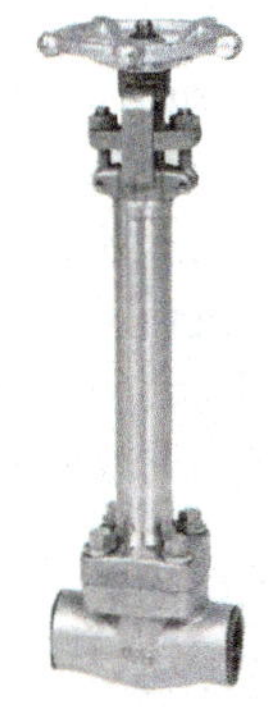

Cryogenic valve

Liquefied gases, such as liquid nitrogen and liquid helium, are used in many cryogenic applications. Liquid nitrogen is the most commonly used element in cryogenics and is legally purchasable around the world. Liquid helium is also commonly used and allows for the lowest attainable temperatures to be reached.

These liquids may be stored in Dewar flasks, which are double-walled containers with a high vacuum between the walls to reduce heat transfer into the liquid. Typical laboratory Dewar flasks are spherical, made of glass and protected in a metal outer container. Dewar flasks for extremely cold liquids such as liquid helium have another double-walled container filled with liquid nitrogen. Dewar flasks are named after their inventor, James Dewar, the man who first liquefied hydrogen. Thermos bottles are smaller vacuum flasks fitted in a protective casing.

Cryogenic barcode labels are used to mark dewar flasks containing these liquids, and will not frost over down to -195 degrees Celsius.

Cryogenic transfer pumps are the pumps used on LNG piers to transfer liquefied natural gas from LNG carriers to LNG storage tanks, as are cryogenic valves.

Cryogenic Processing

The field of cryogenics advanced during World War II when scientists found that metals frozen to low temperatures showed more resistance to wear. Based on this theory of cryogenic hardening, the commercial cryogenic processing industry was founded in 1966 by Ed Busch. With a background in the heat treating industry, Busch founded a company in Detroit called CryoTech in 1966 which merged with 300 Below in 1999 to become the world's largest and oldest commercial cryogenic processing company. Busch originally experimented with the possibility of increasing the life of metal tools to anywhere between 200%-400% of the original life expectancy using cryogenic tempering instead of heat treating. This evolved in the late 1990s into the treatment of other parts.

Cryogens, such as liquid nitrogen, are further used for specialty chilling and freezing applications. Some chemical reactions, like those used to produce the active ingredients for the popular statin drugs, must occur at low temperatures of approximately −100 °C (−148 °F). Special cryogenic chemical reactors are used to remove reaction heat and provide a low temperature environment. The freezing of foods and biotechnology products, like vaccines, requires nitrogen in blast freezing or immersion freezing systems. Certain soft or elastic materials become hard and brittle at very low temperatures, which makes cryogenic milling (cryomilling) an option for some materials that cannot easily be milled at higher temperatures.

Cryogenic processing is not a substitute for heat treatment, but rather an extension of the heating - quenching - tempering cycle. Normally, when an item is quenched, the final temperature is ambient. The only reason for this is that most heat treaters do not have cooling equipment. There is nothing metallurgically significant about ambient temperature. The cryogenic process continues this action from ambient temperature down to −320 °F (140 °R; 78 K; −196 °C). In most instances the cryogenic cycle is followed by a heat tempering procedure. As all alloys do not have the same chemical constituents, the tempering procedure varies according to the material's chemical composition, thermal history and/or a tool's particular service application.

The entire process takes 3–4 days.

Fuels

Another use of cryogenics is cryogenic fuels for rockets with liquid hydrogen as the most widely used example. Liquid oxygen (LOX) is even more widely used but as an oxidizer, not a fuel. NASA's workhorse space shuttle used cryogenic hydrogen/oxygen propellant as its primary means of getting into orbit. LOX is also widely used with RP-1 kerosene, a non-cryogenic hydrocarbon, such as in the rockets built for the Soviet space program by Sergei Korolev.

Russian aircraft manufacturer Tupolev developed a version of its popular design Tu-154 with a cryogenic fuel system, known as the Tu-155. The plane uses a fuel referred to as liquefied natural gas or LNG, and made its first flight in 1989.

Other Applications

Astronomical instruments on the Very Large Telescope are equipped with continuous flow cooling systems.

Some applications of cryogenics:

- Nuclear Magnetic Resonance Spectroscopy (NMR) NMR is one of the most common methods to determine the physical and chemical properties of atoms by detecting the radio frequency absorbed and subsequent relaxation of nuclei in a magnetic field. This is one of the most commonly used characterization techniques and has applications in numerous fields. Primarily, the strong magnetic fields are generated by supercooling electromagnets, although there are spectrometers that do not require cryogens. In traditional superconducting solenoids, liquid helium is used to cool the inner coils because it has a boiling point of around 4 K at ambient pressure. Cheap metallic superconductors can be used for the coil wiring. So-called high-temperature superconducting compounds can be made to super conduct with the use of liquid nitrogen which boils at around 77 K.

- Magnetic resonance imaging (MRI) MRI is a complex application of NMR where the geometry of the resonances is deconvoluted and used to image objects by detecting the relaxation of protons that have been perturbed by a radio-frequency pulse in the strong magnetic field. This is mostly commonly used in health applications.
- Electric power transmission in big cities It is difficult to transmit power by overhead cables in big cities, so underground cables are used. But underground cables get heated and the resistance of the wire increases leading to waste of power. Superconductors could be used to increase power throughput, although they would require cryogenic liquids such as nitrogen or helium to cool special alloy-containing cables to increase power transmission. Several feasibility studies have been performed and the field is the subject of an agreement within the International Energy Agency.

Cryogenic gases delivery truck at a supermarket, Ypsilanti, Michigan

- Frozen food Cryogenic gases are used in transportation of large masses of frozen food. When very large quantities of food must be transported to regions like war zones, earthquake hit regions, etc., they must be stored for a long time, so cryogenic food freezing is used. Cryogenic food freezing is also helpful for large scale food processing industries.
- Forward looking infrared (FLIR) Many infra-red cameras require their detectors to be cryogenically cooled.
- Blood banking Certain rare blood groups are stored at low temperatures, such as −165 °C.
- Special effects Cryogenics technology using liquid nitrogen and CO_2 has been built into nightclub effect systems to create a chilling effect and white fog that can be illuminated with colored lights.
- Manufacturing process Cryogenic cooling is used to cool the tool tip at the time of machining. It increases the tool life. Oxygen is used to perform several important functions in the steel manufacturing process.

- Recycling of Materials By freezing the automobile or truck tire in Liquid nitrogen, the rubber is made brittle & can be crushed into small particles. These particles can be used again for other items.
- Research Experimental research on certain physics phenomena, such as spintronics and magnetotransport properties, requires cryogenic temperatures for the effects to be observed.

Production

Cryogenic cooling of devices and material is usually achieved via the use of liquid nitrogen, liquid helium, or a mechanical cryocooler (which uses high pressure helium lines). Gifford McMahon Cryocoolers, pulse tube cryocoolers and Stirling cryocoolers are in wide use with selection based on required base temperature and cooling capacity. The most recent development in cryogenics is the use of magnets as regenerators as well as refrigerators. These devices work on the principle known as the magnetocaloric effect.

Detectors

For Cryogenic temperature measurement down to 30K, PT-100 sensor (resistance temperature detectors (RTDs)) is used, and for lower than 30K it is required to use Silicon Diode for accuracy. There are also other cryogenic detectors which are used to detect Cryogenic particles.

Cryogenic Deflashing

Cryogenic deflashing is a deflashing process that uses cryogenic temperatures to aid in the removal of flash on cast or molded workpieces. These temperatures cause the flash to become stiff or brittle and to break away cleanly. Cryogenic deflashing is the preferred process when removing excess material from oddly shaped, custom molded products.

Process

Parts are loaded into a parts basket. A cryogen, such as liquid nitrogen, is used to cool the workpieces; once cooled they are tumbled and blasted with media pellets, ranging size from 0.006 to 0.080 inches (0.15 to 2.03 mm). In some instances, cryogenic deflashing does not utilize a blasting action, relying instead only on the tumbling of the parts to remove flash on the outer edges.

Advantages

Cryogenic deflashing provides various advantages over manual deflashing and other traditional deflashing methods.

- The process maintains part integrity and critical tolerances.
- Since it is a batch process, the price per piece is far less as many more parts can be processed in a given amount of time.
- Cryogenic deflashing extends mold life. Rather than replace or repair a mold (which typically involves downtime and high cost), the parts can be deflashed. This is typical of parts molded at the end of their product lifetime.
- The process is computer controlled, therefore removing the human operator variable from the process.
- The process offers consistent results from lot to lot.
- Cryogenic deflashing is non-abrasive.
- The cost per part is generally well below any alternative technique.

Applications

A wide range of molded materials can utilize cryogenic deflashing with proven results. These include:

- Silicones
- Plastics – (both thermoset & thermoplastic)
- Rubbers – (including neoprene & urethane)
- Liquid crystal polymer (LCP)
- Glass-filled nylons
- Aluminum zinc die cast

Examples of applications that use cryogenic deflashing include:

- O-rings & gaskets
- Catheters and other in-vitro medical
- Insulators and other electric / electronic
- Valve stems, washers and fittings
- Tubes and flexible boots
- Face masks & goggles

Today, many molding operations are using cryogenic deflashing instead of rebuilding or repairing molds on products that are approaching their end-of-life. It is often more prudent and economical to add a few cents of production cost for a part than invest in a new molding tool that can cost hundreds of thousands of dollars and has a limited service life due to declining production forecasts.

In other cases, cryogenic deflashing has proven to be an enabling technology, permitting the economical manufacture of high quality, high precision parts fabricated with cutting edge materials and compounds.

Cryogenic Treatment

A cryogenic treatment is the process of treating workpieces to cryogenic temperatures (i.e. below −190 °C (−310 °F)) in order to remove residual stresses and improve wear resistance on steels. In addition to seeking enhanced stress relief and stabilization, or wear resistance, cryogenic treatment is also sought for its ability to improve corrosion resistance by precipitating micro-fine eta carbides, which can be measured before and after in a part using a quantimet.

The process has a wide range of applications from industrial tooling to the improvement of musical signal transmission. Some of the benefits of cryogenic treatment include longer part life, less failure due to cracking, improved thermal properties, better electrical properties including less electrical resistance, reduced coefficient of friction, less creep and walk, improved flatness, and easier machining.

Processes

Cryogenic Hardening

Cryogenic hardening is a cryogenic treatment process where the material is cooled to very low temperatures. By using liquid nitrogen, the temperature can go as low as −190 °C. It can have a profound effect on the mechanical properties of certain materials, such as steels or tungsten carbide.

Applications of Cryogenic Hardening

- Aerospace & Defense: communication, optical housings, weapons platforms, guidance systems, landing systems.
- Automotive: brake rotors, transmissions, clutches, brake parts, rods, crank shafts, camshafts axles, bearings, ring and pinion, heads, valve trains, differentials, springs, nuts, bolts, washers.
- Cutting tools: cutters, knives, blades, drill bits, end mills, turning or milling inserts. There are two main types of cryogenic treatments of cutting tools: Cryogenic treatments of cutting inserts can be classified as follows: Deep Cryo-

genic Treatments (DCT) or Shallow Cryogenic Treatments (SCT). A different minimum tool cooling temperature is used in the two mentioned treatments: -196 °C for DCT and -80 °C for SCT.

- Forming tools: roll form dies, progressive dies, stamping dies.
- Mechanical industry: pumps, motors, nuts, bolts, washers.
- Medical: tooling, scalpels.
- Motorsports and Fleet Vehicles: Automotive for brake rotors and other automotive components.
- Musical: Vacuum tubes, brass instruments, guitar strings and fret wire, piano wire, amplifiers, magnetic pickups, cables, connectors.
- Sports: Firearms, knives, fishing equipment, auto racing, tennis rackets, golf clubs, mountain climbing gear, archery, skiing, aircraft parts, high pressure lines, bicycles, motor cycles.

Cryogenic Machining

Cryogenic machining is a machining process where the traditional flood lubro-cooling liquid (an emulsion of oil into water) is replaced by a jet of either liquid nitrogen (LN2) or pre-compressed carbon dioxide (CO2). Cryogenic machining is useful in rough machining operations, in order to increase the tool life. It can also be useful to preserve the integrity and quality of the machined surfaces in finish machining operations. Cryogenic machining tests have been performed by researchers since several decades, but the actual commercial applications are still limited to very few companies. Both cryogenic machining by turning and milling are possible.

Cryogenic Rolling

Cryogenic rolling or *cryorolling*, is one of the potential techniques to produce nanostructured bulk materials from its bulk counterpart at cryogenic temperatures. It can be defined as rolling that is carried out at cryogenic temperatures. Nanostructured materials are produced chiefly by severe plastic deformation processes. The majority of these methods require large plastic deformations (strains much larger than unity). In case of cryorolling, the deformation in the strain hardened metals is preserved as a result of the suppression of the dynamic recovery. Hence large strains can be maintained and after subsequent annealing, ultra-fine-grained structure can be produced.

Advantages

Comparison of cryorolling and rolling at room temperature:

- In Cryorolling, the strain hardening is retained up to the extent to which rolling is carried out. This implies that there will be no dislocation annihilation and dynamic recovery. Whereas in rolling at room temperature, dynamic recovery is inevitable and softening takes place.
- The flow stress of the material differs for the sample which is subjected to cryorolling. A cryorolled sample has a higher flow stress compared to a sample subjected to rolling at room temperature.
- Cross slip and climb of dislocations are effectively suppressed during cryorolling leading to high dislocation density which is not the case for room temperature rolling.
- The corrosion resistance of the cryorolled sample comparatively decreases due to the high residual stress involved.
- The number of electron scattering centres increases for the cryorolled sample and hence the electrical conductivity decreases significantly.
- The cryorolled sample shows a high dissolution rate.
- Ultra-fine-grained structures can be produced from cryorolled samples after subsequent annealing.

Cryogenic Seal

Cryogenic seals provide a mechanical containment mechanism for materials held at cryogenic temperatures, such as cryogenic fluids. Various techniques, including soldering and welding are available for creating seals; however, specialized materials and processes are necessary to hermetically entrap cryogenic constituents under vacuum-tight conditions. Most commonly used are liquid helium and liquid nitrogen, which boil at very low temperatures, below −153 °C (120 K), as well as hydrocarbons with low freezing points and refrigerating mixtures. Pure indium wire or solder preform washers are accepted as the most reliable low temperature sealing materials. When correctly formed, indium will afford leak rates of less than 4.0x10 -9 mbar- liter/sec. Alternative cryogenic seal materials include silicone grease conical seals, and Pb/Sn (lead-tin) wire seals.

History

Fundamental cryogenic processing began in the 1940s, albeit primitive. Steel cutting tools were immersed in liquid nitrogen to enhance their service life.

Mechanical processes utilizing cryogenics were documented well in the 1950s and by the 1980s cryogenic fluids began to be considered for storage and use in modern devices.

Today, cryogenic seals are a necessity in high-tech commercial, medical, and military applications to encapsulate the cryogenic fluids critical for device resolution and function.

Applications

Applications which utilize cryogenic seals include:

- Magnetic resonance imaging (MRI)
- Chromatography
- Dilution Refrigeration Units
- Cooled Detectors
- Optical Windows
- Infrared Detectors
- Centrifugal Cryogenic Pumps
- Unmanned Aerial Vehicle Systems
- Missile Warning Receivers
- Satellite Tracking Systems
- Infrared Telescopes

Indium Seals

Advantages

Advantages of indium cryogenic seals:

- Established/proven design techniques for indium seal assembly
- Option for disassembly and re-assembly
- Indium can be reformed into useful seals after use
- Soft and pliable at room temperature, due to the low melting temperature of indium, so it fills imperfections. This creates an impervious bond between the mating surfaces, crafting a hermetic seal which remains malleable at cryogenic temperatures
- Seal integrity remains following thermal shock from room temperature to immersion in cryogenic bath

- Seal quality is independent of the mating surface composition, for instance ceramic, germanium, metal, or glass.
- Indium forms a self-passivating oxide layer, 80-100Å thick. This layer is easy to remove with an acid etch, and the underlying, exposed indium metal can be compressed to form a tight, hermetic bond.

Disadvantages

- Bulky mechanical structure required to compress indium between the flanges.
- Pulsating loads cause creep of indium seals, which loosens the bolt tension, thereby reducing the quality of the seal.

Process Information for Indium Seals

- Mating surfaces should be kept as clean as possible, and may be cleaned using acetone.
- Clean, oxide-free indium will cold weld to itself. The mating ends of a wire seal will weld together under compression.
- A more reliable alternative to a seal made from indium wire is a seal that uses an indium washer. Washers minimize the risk of seal degradation and cryogenic leaks by eliminating the interface between connected butt ends of wire. Washers are manufactured as a continuous ring with no breaks.
- As many fasteners as possible should be used to clamp the indium seal.
- Indium material used must be ultra-pure (99.9 minimum purity) to prevent hardening of the material at sub-zero temperatures, as well as to restrict impurities of elements with low vapor-pressure.
- Material used for indium cryogenic seals should be manufactured from vacuum-cast material to prevent outgassing after fixturing in the assembly.

Reliability Testing

- Helium leak tests
- Cryogenic temperature shock testing

Types

- Compact indium seal
- Compressible hermetic seal

- Compression seal
- Cryogenic vacuum seal
- Demountable cryogenic seal
- Indium cryogenic vacuum seal
- Indium o-ring flange seals
- Indium seal
- Indium wire o-ring
- Indium wire seal
- Low profile indium seal
- Low temperature seal
- Reusable cryogenic vacuum seal
- Reusable indium wire seal
- Reusable, low-profile, cryogenic wire seal
- Soft metal seal
- Vacuum compatible seal
- Vacuum compatible seal at cryogenic temperature
- Corner joint seal
- Face joint seal

Manufacturers

- Indium Corporation
- CMR-Direct
- Garlock Helicoflex
- AIM
- ESPI Metals
- Sealwise
- Kapton®

- Mylar®
- Indium Wire Extrusion
- VIDEO showing how to make an indium seal on a cryogenic system

Cryogenic Fuel

Cryogenic fuels are fuels that require storage at extremely low temperatures in order to maintain them in a liquid state. These fuels are used in machinery that operates in space (e.g. rocket ships and satellites) because ordinary fuel cannot be used there, due to absence of an environment that supports combustion (on earth, oxygen is abundant in the atmosphere, whereas in human-explorable space, oxygen is virtually non-existent). Cryogenic fuels most often constitute liquefied gases such as liquid hydrogen.

Some rocket engines use regenerative cooling, the practice of circulating their cryogenic fuel around the nozzles before the fuel is pumped into the combustion chamber and ignited. This arrangement was first suggested by Eugen Sänger in the 1940s. The Saturn V rocket that sent the first manned missions to the moon used this design element, which is still in use today.

Quite often, liquid oxygen is mistakenly called cryogenic "fuel", though it is actually an oxidizer and not a fuel.

Russian aircraft manufacturer Tupolev developed a version of its popular Tu-154 design but with a cryogenic fuel system, designated the Tu-155. Using a fuel referred to as liquefied natural gas (LNG), its first flight was in 1989.

Operation

Cryogenic fuels can be placed into two categories: inert and flammable or combustible. Both types exploit the large liquid to gas volume ratio that occurs when liquid transitions to gas phase. The feasibility of cryogenic fuels is associated with what is known as a high mass flow rate. With regulation, the high-density energy of cryogenic fuels is utilized to produce thrust in rockets and controllable consumption of fuel. The following sections provide further detail.

Inert

These types of fuels typically use the regulation of gas production and flow to power pistons in an engine. The large increases in pressure are controlled and directed toward the engine's pistons. The pistons move due to the mechanical power transformed from the monitored production of gaseous fuel. A notable example can be in Peter Dear-man's liquid air vehicle. Some common inert fuels include:

- Liquid nitrogen
- Liquid air
- Liquid helium
- Liquid neon

Combustible

These fuels utilize the beneficial liquid cryogenic properties along with the flammable nature of the substance as a source of power. These types of fuel are well known primarily for their use in rockets including the Intercontinental ballistic missile. Some common combustible fuels include:

- Liquid hydrogen
- Liquid natural gas (LNG)
- Liquid methane

Engine Combustion

Combustible cryogenic fuels offer much more utility than most inert fuels can by. Liquefied natural gas, as with any fuel, will only combust when properly mixed with right amounts of air. As for LNG, the bulk majority of efficiency depends on the methane number, which is the gas equivalent of the octane number. This is determined based on the methane content of the liquefied fuel and any other dissolved gas, and varies as a result of experimental efficiencies. Maximizing efficiency in combustion engines will be a result of determining the proper fuel to air ratio and utilizing the addition other hydrocarbons for added optimal combustion.

Production Efficiency

Current gas liquefying processes have been improving over the past decades with the advent of better machinery and control of system heat losses. Typical techniques take advantage of the temperature of the gas dramatically cooling as the controlled pressure of a gas is released. Enough pressurization and then subsequent depressurization can liquefy most gases, first noticed via the Joule-Thomson effect.

Liquefied Natural Gas

While it is cost effective to liquefy natural gas for storage, transport, and use, roughly 10 to 15 percent of the gas gets consumed during the process. The optimal process contains four stages of propane refrigeration and two stages of ethylene refrigeration. There can be the addition of an additional refrigerant stage, but the additional costs of equipment are not economically justifiable. Efficiency can be tied to the pure component cascade processes which minimize the overall source to sink temperature difference associated with refrigerant condensing. The optimized process incorporates optimized heat recovery along with the use of pure refrigerants. All process designers of liquefaction plants using proven technologies face the same challenge: to efficiently cool and condense a mixture with a pure refrigerant. In the optimized Cascade process, the mixture to be cooled and condensed is the feed gas. In the propane mixed refrigerant processes, the two mixtures requiring cooling and condensing are the feed gas and the mixed refrigerant. The chief source of inefficiency lies in the heat exchange train during the liquefaction process.

Advantages and Disadvantages

Benefits

- Cryogenic fuels are environmentally cleaner than gasoline or fossil fuels. Among other things, the greenhouse gas rate could potentially be reduced by 11%-20% using LNG as opposed to gasoline when transporting goods.
- Along with their eco-friendly nature, they have the potential to significantly decrease transportation costs of inland products because of their abundance compared to that of fossil fuels.
- Cryogenic fuels have a higher mass flow rate than fossil fuels and therefore produce more thrust and power when combusted for use in an engine. This means that engines will run farther on less fuel overall than modern gas engines.
- Cryogenic fuels are non-pollutants and therefore, if spilled, are no risk to the environment. There will be no need to clean up hazardous waste after a spill.

Potential Drawbacks

- Some cryogenic fuels, like LNG, are naturally combustible. Ignition of fuel spills could result in a large explosion. This is possible in the case of a car crash with an LNG engine.
- Cryogenic storage tanks must be able to withstand high pressure. High-pressure propellant tanks require thicker walls and stronger alloys which make the vehicle tanks heavier, thereby reducing performance and practicality.
- Despite non-toxic tendencies, cryogenic fuels are denser than air. As such, they can lead to asphyxiation. If leaked, the liquid will boil into a very dense, cold gas and if inhaled, could be fatal.

Cryogenic Energy Storage

Cryogenic energy storage (CES) is the use of low temperature (cryogenic) liquids such as liquid air or liquid nitrogen as energy storage. Both cryogens have been used to power cars. The inventor Peter Dearman initially developed a liquid air car, and then used the technology he developed for grid energy storage. The technology is being piloted at a UK power station.

History

A liquid air powered car called Liquid Air was built between 1899 and 1902 but it couldn't at the time compete in terms of efficiency with other engines. More recently, a liquid nitrogen vehicle was built. Peter Dearman, a garage inventor in Hertfordshire, UK who had initially developed a liquid air powered car, then put the technology to use as grid energy storage. The Dearman engine differs from former nitrogen engine designs in that the nitrogen is heated by combining it with the heat exchange fluid inside the cylinder of the engine.

Grid Energy Storage

Process

When it is cheaper (usually at night), electricity is used to cool air from the atmosphere to -195 °C using the Claude Cycle to the point where it liquefies. The liquid air, which takes up one-thousandth of the volume of the gas, can be kept for a long time in a large vacuum flask at atmospheric pressure. At times of high demand for electricity, the liquid air is pumped at high pressure into a heat exchanger, which acts as a boiler. Air from the atmosphere at ambient temperature, or hot water from an industrial heat source, is used to heat the liquid and turn it back into a gas. The massive increase in volume and pressure from this is used to drive a turbine to generate electricity.

Efficiency

In isolation the process is only 25% efficient, but this is greatly increased (to around 50%) when used with a low-grade cold store, such as a large gravel bed, to capture the cold generated by evaporating the cryogen. The cold is re-used during the next refrigeration cycle.

Efficiency is further increased when used in conjunction with a power plant or other source of low-grade heat that would otherwise be lost to the atmosphere. Highview Power Storage claims an AC to AC round-trip efficiency of 70%, by using an otherwise waste heat source at 115 °C. The IMechE (Institution of Mechanical Engineers) agrees that these estimates for a commercial-scale plant are realistic. However this number was not checked or confirmed by independent professional institutions.

Currently surplus gaseous nitrogen is produced as a byproduct in the production of oxygen. Oxygen can be used in oxy-combustion coal power plants, enabling CO_2 capture and sequestration. This gaseous nitrogen can be liquefied by available liquefaction capacities for further use. Cryogenic distillation of air is currently the only commercially viable technology for large scale oxygen production.

Pilot Plant

A 300 kW, 2.5MWh storage capacity pilot cryogenic energy system developed by researchers at the University of Leeds and Highview Power Storage, that uses liquid air (with the CO_2 and water removed as they would turn solid at the storage temperature) as the energy store, and low-grade waste heat to boost the thermal re-expansion of the air, has been operating at a 80MW biomass power station in Slough, UK, since 2010. the efficiency is less than 15% because of low efficiency hardware components used, but the engineers are targeting an efficiency of about 60 percent for the next generation of CES based on operation experiences of this system.

The system is based on proven technology, used safely in many industrial processes, and does not require any particularly rare elements or expensive components to manufacture. Dr Tim Fox, the head of Energy at the IMechE says "it uses standard industrial components...., it will last for decades, and it can be fixed with a spanner."

In April 2014 the UK government announced it had given them £8 million to fund the next stage of the demonstration

Cryogenic Storage Dewar

A cryogenic storage dewar is a specialised type of vacuum flask used for storing cryogens (such as liquid nitrogen or liquid helium), whose boiling points are much lower than room temperature. Cryogenic storage dewars may take several different forms including open buckets, flasks with loose-fitting stoppers and self-pressurising tanks. All dewars have walls constructed from two or more layers, with a high vacuum maintained between the layers. This provides very good thermal insulation between the interior and exterior of the dewar, which reduces the rate at which the contents boil away. Precautions are taken in the design of dewars to safely manage the gas which is released as the liquid slowly boils. The simplest dewars allow the gas to escape either through an open top or past a loose-fitting stopper to prevent the risk of explosion. More sophisticated dewars trap the gas above the liquid, and hold it at high pressure. This increases the boiling point of the liquid, allowing it to be stored for extended periods. Excessive vapour pressure is released automatically through safety valves. The method of decanting liquid from a dewar depends upon its design. Simple dewars may be tilted, to pour liquid from the neck. Self-pressurising designs use the gas pressure in the top of the dewar to force the liquid upward through a pipe leading to the neck.

Safety

Cryogens present several safety hazards and storage vessels are designed to reduce the associated risk. Firstly, no dewar can provide perfect thermal insulation and the cryogenic liquid slowly boils away, which yields an enormous quantity of gas. For example, the expansion ratio of cryogenic argon from the boiling point to ambient is 1 to 847, liquid hydrogen 1 to 851, liquid helium 1 to 757, liquid nitrogen 1 to 696, and liquid oxygen 1 to 860; Neon has the highest expansion ratio with 1 to 1438. In dewars with an open top, the gas simply escapes into the surrounding area. However, very high pressures can build up inside sealed dewars, and precautions are taken to minimise the risk of explosion. One or more pressure-relief valves allow gas to vent away from the dewar whenever the pressure becomes excessively large. In an incident in 2006 at Texas A&M University, the pressure-relief devices of a tank of liquid nitrogen were sealed with brass plugs. As a result, the tank failed catastrophically and exploded.

Ice can form on the inside of the dewar if it is left open to the air for extended periods. This can be extremely dangerous as the openings of the dewar can become blocked, leading to a pressure build-up, and the risk of explosion.

The gas escaping from a dewar can gradually displace the oxygen from the air in the surrounding area, which presents an asphyxiation hazard. Users are trained to only store dewars in a well-ventilated area and before transporting dewars in an elevator, the excess gas pressure is vented away and the dewars are sent unaccompanied to their destination.

Crystal

A crystal or crystalline solid is a solid material whose constituents (such as atoms, molecules or ions) are arranged in a highly ordered microscopic structure, forming a crystal lattice that extends in all directions. In addition, macroscopic single crystals are usually identifiable by their geometrical shape, consisting of flat faces with specific, characteristic orientations. The scientific study of crystals and crystal formation is known as crystallography. The process of crystal formation via mechanisms of crystal growth is called crystallization or solidification.

A crystal of amethyst quartz

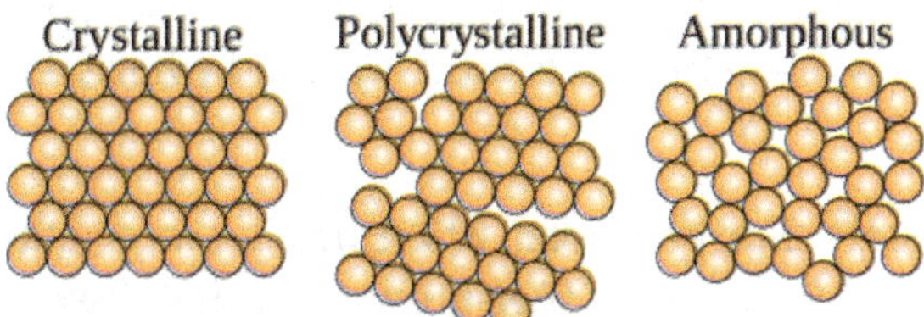

Microscopically, a single crystal has atoms in a near-perfect periodic arrangement; a polycrystal is composed of many microscopic crystals (called "crystallites" or "grains"); and an amorphous solid (such as glass) has no periodic arrangement even microscopically.

The word *crystal* derives from the Ancient Greek word *krustallos* meaning both "ice" and "rock crystal", from *kruos* "icy cold, frost".

Examples of large crystals include snowflakes, diamonds, and table salt. Most inorganic solids are not crystals but polycrystals, i.e. many microscopic crystals fused together into a single solid. Examples of polycrystals include most metals, rocks, ceramics, and ice. A third category of solids is amorphous solids, where the atoms have no periodic structure whatsoever. Examples of amorphous solids include glass, wax, and many plastics.

Crystal Structure (Microscopic)

Halite (table salt, NaCl): Microscopic and macroscopic

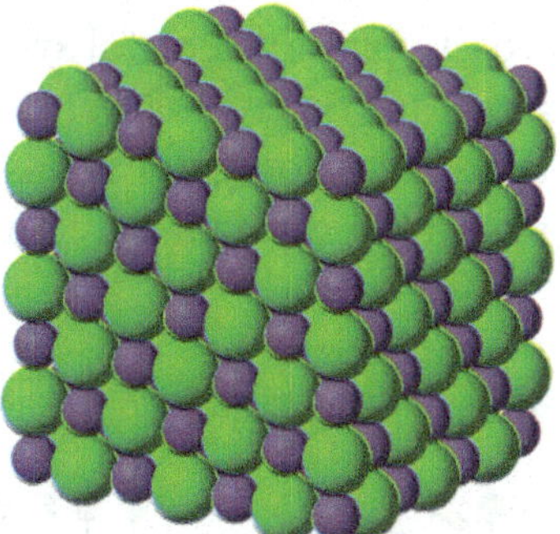

Microscopic structure of a halite crystal. (Purple is sodium ion, green is chlorine ion.) There is cubic symmetry in the atoms' arrangement.

Macroscopic (~16cm) halite crystal. The right-angles between crystal faces are due to the cubic symmetry of the atoms' arrangement.

The scientific definition of a "crystal" is based on the microscopic arrangement of atoms inside it, called the crystal structure. A crystal is a solid where the atoms form a periodic arrangement.

Not all solids are crystals. For example, when liquid water starts freezing, the phase change begins with small ice crystals that grow until they fuse, forming a *polycrystalline* structure. In the final block of ice, each of the small crystals (called "crystallites" or "grains") is a true crystal with a periodic arrangement of atoms, but the whole polycrystal does *not* have a periodic arrangement of atoms, because the periodic pattern is broken at the grain boundaries. Most macroscopic inorganic solids are polycrystalline, including almost all metals, ceramics, ice, rocks, etc. Solids that are neither crystalline nor polycrystalline, such as glass, are called *amorphous solids*, also called glassy, vitreous, or noncrystalline. These have no periodic order, even microscopically. There are distinct differences between crystalline solids and amorphous solids: most notably, the process of forming a glass does not release the latent heat of fusion, but forming a crystal does.

A crystal structure (an arrangement of atoms in a crystal) is characterized by its *unit cell*, a small imaginary box containing one or more atoms in a specific spatial arrangement. The unit cells are stacked in three-dimensional space to form the crystal.

The symmetry of a crystal is constrained by the requirement that the unit cells stack perfectly with no gaps. There are 219 possible crystal symmetries, called crystallographic space groups. These are grouped into 7 crystal systems, such as cubic crystal system (where the crystals may form cubes or rectangular boxes, such as halite shown at right) or hexagonal crystal system (where the crystals may form hexagons, such as ordinary water ice).

Crystal Faces and Shapes

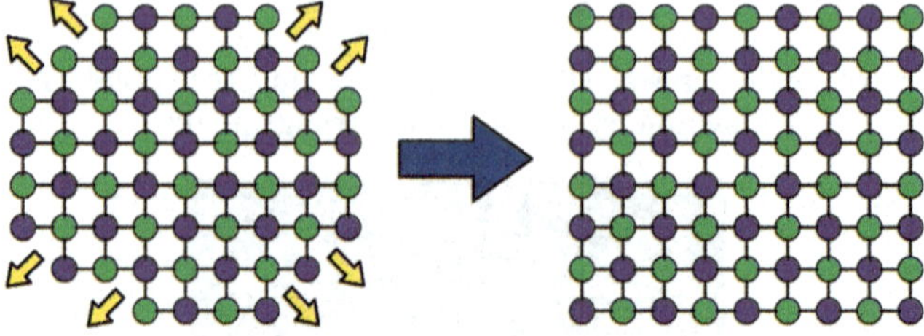

As a halite crystal is growing, new atoms can very easily attach to the parts of the surface with rough atomic-scale structure and many dangling bonds. Therefore, these parts of the crystal grow out very quickly (yellow arrows). Eventually, the whole surface consists of smooth, stable faces, where new atoms cannot as easily attach themselves.

Crystals are commonly recognized by their shape, consisting of flat faces with sharp angles. These shape characteristics are not *necessary* for a crystal—a crystal is scientifically defined by its microscopic atomic arrangement, not its macroscopic shape—but the characteristic macroscopic shape is often present and easy.

Euhedral crystals are those with obvious, well-formed flat faces. Anhedral crystals do not, usually because the crystal is one grain in a polycrystalline solid.

The flat faces (also called facets) of a euhedral crystal are oriented in a specific way relative to the underlying atomic arrangement of the crystal: they are planes of relatively low Miller index. This occurs because some surface orientations are more stable than others (lower surface energy). As a crystal grows, new atoms attach easily to the rougher and less stable parts of the surface, but less easily to the flat, stable surfaces. Therefore, the flat surfaces tend to grow larger and smoother, until the whole crystal surface consists of these plane surfaces.

One of the oldest techniques in the science of crystallography consists of measuring the three-dimensional orientations of the faces of a crystal, and using them to infer the underlying crystal symmetry.

A crystal's habit is its visible external shape. This is determined by the crystal structure (which restricts the possible facet orientations), the specific crystal chemistry and bonding (which may favor some facet types over others), and the conditions under which the crystal formed.

Occurrence in Nature

Ice crystals.

Fossil shell with calcite crystals.

Rocks

By volume and weight, the largest concentrations of crystals in the Earth are part of its solid bedrock. Crystals found in rocks typically range in size from a fraction of a millimetre to several centimetres across, although exceptionally large crystals are occasionally found. As of 1999, the world's largest known naturally occurring crystal is a crystal of beryl from Malakialina, Madagascar, 18 m (59 ft) long and 3.5 m (11 ft) in diameter, and weighing 380,000 kg (840,000 lb).

Some crystals have formed by magmatic and metamorphic processes, giving origin to large masses of crystalline rock. The vast majority of igneous rocks are formed from molten magma and the degree of crystallization depends primarily on the conditions under which they solidified. Such rocks as granite, which have cooled very slowly and under great pressures, have completely crystallized; but many kinds of lava were poured out at the surface and cooled very rapidly, and in this latter group a small amount of amorphous or glassy matter is common. Other crystalline rocks, the metamorphic rocks such as marbles, mica-schists and quartzites, are recrystallized. This means that they were at first fragmental rocks like limestone, shale and sandstone and have never been in a molten condition nor entirely in solution, but the high temperature and pressure conditions of metamorphism have acted on them by erasing their original structures and inducing recrystallization in the solid state.

Other rock crystals have formed out of precipitation from fluids, commonly water, to form druses or quartz veins. The evaporites such as halite, gypsum and some limestones have been deposited from aqueous solution, mostly owing to evaporation in arid climates.

Ice

Water-based ice in the form of snow, sea ice and glaciers is a very common manifestation of crystalline or polycrystalline matter on Earth. A single snowflake is typically a single crystal, while an ice cube is a polycrystal.

Organigenic Crystals

Many living organisms are able to produce crystals, for example calcite and aragonite in the case of most molluscs or hydroxylapatite in the case of vertebrates.

Polymorphism and Allotropy

The same group of atoms can often solidify in many different ways. Polymorphism is the ability of a solid to exist in more than one crystal form. For example, water ice is ordinarily found in the hexagonal form Ice I_h, but can also exist as the cubic Ice I_c, the rhombohedral ice II, and many other forms. The different polymorphs are usually called different *phases*.

In addition, the same atoms may be able to form noncrystalline phases. For example, water can also form amorphous ice, while SiO_2 can form both fused silica (an amorphous glass) and quartz (a crystal). Likewise, if a substance can form crystals, it can also form polycrystals.

For pure chemical elements, polymorphism is known as allotropy. For example, diamond and graphite are two crystalline forms of carbon, while amorphous carbon is a noncrystalline form. Polymorphs, despite having the same atoms, may have wildly different properties. For example, diamond is among the hardest substances known, while graphite is so soft that it is used as a lubricant.

Polyamorphism is a similar phenomenon where the same atoms can exist in more than one amorphous solid form.

Crystallization

Vertical cooling crystallizer in a beet sugar factory.

Crystallization is the process of forming a crystalline structure from a fluid or from materials dissolved in a fluid. (More rarely, crystals may be deposited directly from gas; thin-film deposition and epitaxy.)

Crystallization is a complex and extensively-studied field, because depending on the conditions, a single fluid can solidify into many different possible forms. It can form a single crystal, perhaps with various possible phases, stoichiometries, impurities, defects, and habits. Or, it can form a polycrystal, with various possibilities for the size, arrangement, orientation, and phase of its grains. The final form of the solid is determined by the conditions under which the fluid is being solidified, such as the chemistry of the fluid, the ambient pressure, the temperature, and the speed with which all these parameters are changing.

Specific industrial techniques to produce large single crystals (called *boules*) include the Czochralski process and the Bridgman technique. Other less exotic methods of crystallization may be used, depending on the physical properties of the substance, including hydrothermal synthesis, sublimation, or simply solvent-based crystallization.

Large single crystals can be created by geological processes. For example, selenite crystals in excess of 10 meters are found in the Cave of the Crystals in Naica, Mexico. For more details on geological crystal formation.

Crystals can also be formed by biological processes, Conversely, some or-ganisms have special techniques to *prevent* crystallization from occurring, such as an-tifreeze proteins.

Defects, Impurities, and Twinning

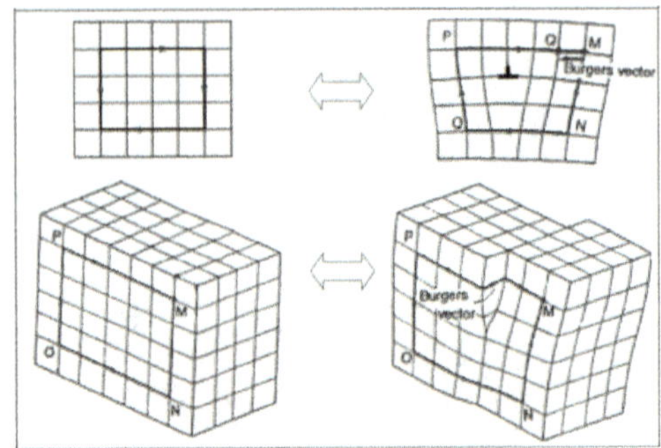

Two types of crystallographic defects. Top right: edge dislocation. Bottom right: screw dislocation.

An *ideal* crystal has every atom in a perfect, exactly repeating pattern. However, in reality, most crystalline materials have a variety of crystallographic defects, places where the crystal's pattern is interrupted. The types and structures of these defects may have a profound effect on the properties of the materials.

A few examples of crystallographic defects include vacancy defects (an empty space where an atom should fit), interstitial defects (an extra atom squeezed in where it does not fit), and dislocations . Dislocations are especially im-portant in materials science, because they help determine the mechanical strength of materials.

Another common type of crystallographic defect is an impurity, meaning that the "wrong" type of atom is present in a crystal. For example, a perfect crystal of diamond would only contain carbon atoms, but a real crystal might perhaps contain a few boron atoms as well. These boron impurities change the diamond's color to slightly blue. Likewise, the only difference between ruby and sapphire is the type of impurities present in a corundum crystal.

In semiconductors, a special type of impurity, called a dopant, drastically changes the crystal's electrical properties. Semiconductor devices, such as transistors, are made possible largely by putting different semiconductor dopants into different places, in specific patterns.

Twinned pyrite crystal group.

Twinning is a phenomenon somewhere between a crystallographic defect and a grain boundary. Like a grain boundary, a twin boundary has different crystal orientations on its two sides. But unlike a grain boundary, the orientations are not random, but related in a specific, mirror-image way.

Mosaicity is a spread of crystal plane orientations. A mosaic crystal is supposed to consist of smaller crystalline units that are somewhat misaligned with respect to each other.

Chemical Bonds

In general, solids can be held together by various types of chemical bonds, such as metallic bonds, ionic bonds, covalent bonds, van der Waals bonds, and others. None of these are necessarily crystalline or non-crystalline. However, there are some general trends as follows.

Metals are almost always polycrystalline, though there are exceptions like amorphous metal and single-crystal metals. The latter are grown synthetically. (A microscopically-small piece of metal may naturally form into a single crystal, but larger pieces generally do not.) Ionically bonded solids are usually crystalline or polycrystalline. In practice, large salt crystals can be created by solidification of a molten fluid, or by crystallization out of a solution. Covalently bonded crystals are also very common, notable examples being diamond, quartz, and graphite. Polymer materials generally will form crystalline regions, but the lengths of the molecules usually prevent complete crystallization—and sometimes polymers are completely amorphous. Weak van der Waals forces also help hold together certain crystals, including graphite.

Quasicrystals

The material holmium–magnesium–zinc (Ho–Mg–Zn) forms quasicrystals, which can take on the macroscopic shape of a dodecahedron. (Only a quasicrystal, not a normal crystal, can take this shape.) The edges are 2 mm long.

A quasicrystal consists of arrays of atoms that are ordered but not strictly periodic. They have many attributes in common with ordinary crystals, such as displaying a discrete pattern in x-ray diffraction, and the ability to form shapes with smooth, flat faces.

Quasicrystals are most famous for their ability to show five-fold symmetry, which is impossible for an ordinary periodic crystal.

The International Union of Crystallography has redefined the term "crystal" to include both ordinary periodic crystals and quasicrystals ("any solid having an essentially discrete diffraction diagram").

Quasicrystals, first discovered in 1982, are quite rare in practice. Only about 100 solids are known to form quasicrystals, compared to about 400,000 periodic crystals known in 2004. The 2011 Nobel Prize in Chemistry was awarded to Dan Shechtman for the discovery of quasicrystals.

Special Properties from Anisotropy

Crystals can have certain special electrical, optical, and mechanical properties that glass and polycrystals normally cannot. These properties are related to the anisotropy of the crystal, i.e. the lack of rotational symmetry in its atomic arrangement. One such property is the piezoelectric effect, where a voltage across the crystal can shrink or stretch it. Another is birefringence, where a double image appears when looking through a crystal. Moreover, various properties of a crystal, including electrical conductivity, electrical permittivity, and Young's modulus, may be different in different directions in a crystal. For example, graphite crystals consist of a stack of sheets, and although each individual sheet is mechanically very strong, the sheets are rather loosely bound to each other. Therefore, the mechanical strength of the material is quite different depending on the direction of stress.

Not all crystals have all of these properties. Conversely, these properties are not quite exclusive to crystals. They can appear in glasses or polycrystals that have been made anisotropic by working or stress—for example, stress-induced birefringence.

Crystallography

Crystallography is the science of measuring the crystal structure (in other words, the atomic arrangement) of a crystal. One widely used crystallography technique is X-ray diffraction. Large numbers of known crystal structures are stored in crystallographic databases.

Gallery

Insulin crystals grown in earth orbit.

An apatite crystal sits front and center on cherry-red rhodochroite rhombs, purple fluorite cubes, quartz and a dusting of brass-yellow pyrite cubes.

Cryotank

Cryotank or cryogenic tank is a tank that is used to store frozen biological material.

Cryotanks and cryogenics can be seen in many sci-fi movies but today, it is still an up and rising technology. The term "cryotank" refers to storage of super-cold fuels, such as liquid oxygen and liquid hydrogen. The premise of it seems very simple. All that needs to be done is for a human to be loaded into the tank and then they can be frozen until a time comes when any diseases they have can be cured and they can live an even longer life. This could also be used in space travel and just preserving human life in general. The problem with this is when the human body is frozen, ice crystals form in the cells. The ice crystals then continue to expand rupturing the cell wall and destroying the integrity of the cell, or killing it.

Cryotanks at the Institute of Plant Industry, Saint Petersburg, Russia.

This means in order for humans to undergo the cryogenic process a way to significantly raise the levels of glucose produced in the human body is needed.

In Fiction

Cryogenic Tanks are used to store natural gases such as oxygen, argon, nitrogen, helium and other materials. By putting them in these tanks they can be stored at the correct temperature and pressure to be transported to different areas that need them for use.

In Science Fiction

In science fiction, it is used to freeze people. (*cry-* is a Greek prefix which means "cold or freezing") So people are stored in the tank frozen to come out in the future. Cryotanks are found in some science fiction films such as *Prometheus* (2012) and *The Host* (2013).

Absolute zero

Absolute zero is the lower limit of the thermodynamic temperature scale, a state at which the enthalpy and entropy of a cooled ideal gas reaches its minimum value, taken as 0. The theoretical temperature is determined by extrapolating the ideal gas law; by international agreement, absolute zero is taken as −273.15° on the Celsius scale (International System of Units), which equates to −459.67° on the Fahrenheit scale (United States customary units or Imperial units). The corresponding Kelvin and Rankine temperature scales set their zero points at absolute zero by definition.

It is commonly thought of as the lowest temperature possible, but it is not the lowest *enthalpy* state possible, because all real substances begin to depart from the ideal gas when cooled as they approach the change of state to liquid, and then to solid; and the sum of the enthalpy of vaporization (gas to liquid) and enthalpy of fusion (liquid to solid) exceeds the ideal gas's change in enthalpy to absolute zero. In the quantum-mechanical description, matter (solid) at absolute zero is in its ground state, the point of lowest internal energy.

The laws of thermodynamics indicate that absolute zero cannot be reached using only thermodynamic means, because the temperature of the substance being cooled approaches the temperature of the cooling agent asymptotically. And a system at absolute zero still possesses quantum mechanical zero-point energy, the energy of its ground state at absolute zero. The kinetic energy of the ground state cannot be removed.

Scientists and technologists routinely achieve temperatures close to absolute zero, where matter exhibits quantum effects such as superconductivity and superfluidity.

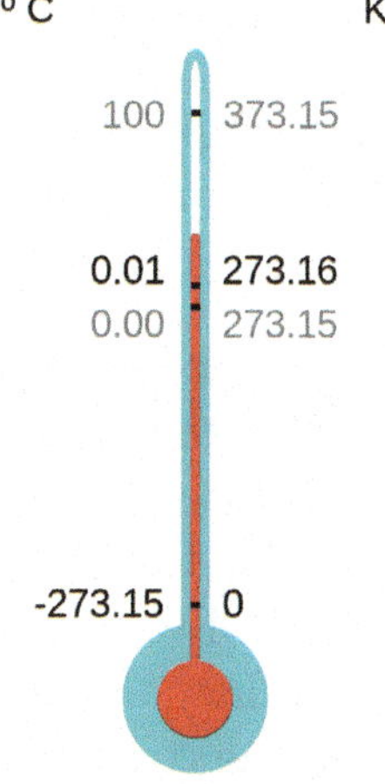

Zero kelvin (−273.15°C) is defined to be absolute zero.

Thermodynamics Near Absolute Zero

At temperatures near 0 K, nearly all molecular motion ceases and $\Delta S = 0$ for any adiabatic process, where S is the entropy. In such a circumstance, pure substances can (ideally) form perfect crystals as $T \to 0$. Max Planck's strong form of the third law of thermodynamics states the entropy of a perfect crystal vanishes at absolute zero. The original Nernst *heat theorem* makes the weaker and less controversial claim that the entropy change for any isothermal process approaches zero as $T \to 0$:

$$\lim_{T \to 0} \Delta S = 0$$

The implication is that the entropy of a perfect crystal simply approaches a constant value.

The Nernst postulate identifies the isotherm T = 0 as coincident with the adiabat S = 0, although other isotherms and adiabats are distinct. As no two adiabats intersect, no other adiabat can intersect the T = 0 isotherm. Consequently no adiabatic process initiated at nonzero temperature can lead to zero temperature. *(≈ Callen, pp. 189–190)*

An even stronger assertion is that *It is impossible by any procedure to reduce the temperature of a system to zero in a finite number of operations.* (≈ Guggenheim, p. 157)

A perfect crystal is one in which the internal lattice structure extends uninterrupted in all directions. The perfect order can be represented by translational symmetry along three (not usually orthogonal) axes. Every lattice element of the structure is in its proper place, whether it is a single atom or a molecular grouping. For substances which have two (or more) stable crystalline forms, such as diamond and graphite for carbon, there is a kind of "chemical degeneracy". The question remains whether both can have zero entropy at $T = 0$ even though each is perfectly ordered.

Perfect crystals never occur in practice; imperfections, and even entire amorphous material inclusions, can and do simply get "frozen in" at low temperatures, so transitions to more stable states do not occur.

Using the Debye model, the specific heat and entropy of a pure crystal are proportional to T^3, while the enthalpy and chemical potential are proportional to T^4. (Guggenheim, p. 111) These quantities drop toward their $T = 0$ limiting values and approach with *zero* slopes. For the specific heats at least, the limiting value itself is definitely zero, as borne out by experiments to below 10 K. Even the less detailed Einstein model shows this curious drop in specific heats. In fact, all specific heats vanish at absolute zero, not just those of crystals. Likewise for the coefficient of thermal expansion. Maxwell's relations show that various other quantities also vanish. These phenomena were unanticipated.

Since the relation between changes in Gibbs free energy (G), the enthalpy (H) and the entropy is

$$\Delta G = \Delta H - T\Delta S$$

thus, as T decreases, ΔG and ΔH approach each other (so long as ΔS is bounded). Experimentally, it is found that all spontaneous processes (including chemical reactions) result in a decrease in G as they proceed toward equilibrium. If ΔS and/or T are small, the condition $\Delta G < 0$ may imply that $\Delta H < 0$, which would indicate an exothermic reaction. However, this is not required; endothermic reactions can proceed spontaneously if the $T\Delta S$ term is large enough.

Moreover, the slopes of the derivatives of ΔG and ΔH converge and are equal to zero at $T = 0$. This ensures that ΔG and ΔH are nearly the same over a considerable range of temperatures and justifies the approximate empirical Principle of Thomsen and Berthelot, which states that *the equilibrium state to which a system proceeds is the one which evolves the greatest amount of heat*, i.e. an actual process is the *most exothermic one*. (Callen, pp. 186–187)

One model that estimates the properties of an electron gas at absolute zero in metals is the Fermi gas. The electrons, being Fermions, have to be in different quantum states, which leads the electrons to get very high typical velocities, even at absolute zero. The maximum energy that electrons can have at absolute zero is called the Fermi energy. The Fermi temperature is defined as this maximum energy divided by Boltzmann's constant, and is of the order of 80,000 K for typical electron densities found in metals. For temperatures significantly below the Fermi temperature, the electrons behave in almost the same way as at absolute zero. This explains the failure of the classical equipartition theorem for metals that eluded classical physicists in the late 19th century.

Relation with Bose–Einstein Condensates

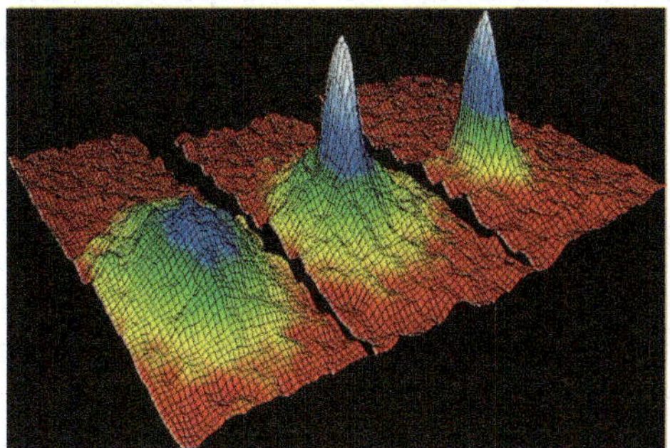

Velocity-distribution data of a gas of rubidium atoms at a temperature within a few billionths of a degree above absolute zero. Left: just before the appearance of a Bose–Einstein condensate. Center: just after the appearance of the condensate. Right: after further evaporation, leaving a sample of nearly pure condensate.

A Bose–Einstein condensate (BEC) is a state of matter of a dilute gas of weakly interacting bosons confined in an external potential and cooled to temperatures very near absolute zero. Under such conditions, a large fraction of the bosons occupy the lowest quantum state of the external potential, at which point quantum effects become apparent on a macroscopic scale.

This state of matter was first predicted by Satyendra Nath Bose and Albert Einstein in 1924–25. Bose first sent a paper to Einstein on the quantum statistics of light quanta (now called photons). Einstein was impressed, translated the paper from English to German and submitted it for Bose to the *Zeitschrift für Physik* which published it. Einstein then extended Bose's ideas to material particles (or matter) in two other papers.

Seventy years later, in 1995, the first gaseous condensate was produced by Eric Cornell and Carl Wieman at the University of Colorado at Boulder NIST-JILA lab, using a gas of rubidium atoms cooled to 170 nanokelvin (nK) (1.7×10^{-7} K).

A record cold temperature of 450 ±80 pK in a Bose–Einstein condensate (BEC) of sodium atoms was achieved in 2003 by researchers at MIT. The associated blackbody (peak emittance) wavelength of 6,400 kilometers is roughly the radius of Earth.

Absolute Temperature Scales

Absolute, or thermodynamic, temperature is conventionally measured in kelvins (Celsius-scaled increments) and in the Rankine scale (Fahrenheit-scaled increments) with increasing rarity. Absolute temperature measurement is uniquely determined by a multiplicative constant which specifies the size of the "degree", so the *ratios* of two absolute temperatures, T_2/T_1, are the same in all scales. The most transparent definition of this standard comes from the Maxwell–Boltzmann distribution. It can also be found in Fermi–Dirac statistics (for particles of half-integer spin) and Bose–Einstein statistics (for particles of integer spin). All of these define the relative numbers of particles in a system as decreasing exponential functions of energy (at the particle level) over kT, with k representing the Boltzmann constant and T representing the temperature observed at the macroscopic level.

Negative Temperatures

Temperatures that are expressed as negative numbers on the familiar Celsius or Fahrenheit scales are simply colder than the zero points of those scales. Certain systems can achieve truly negative temperatures; that is, their thermodynamic temperature (expressed in kelvin) can be of a negative quantity. A system with a truly negative temperature is not colder than absolute zero. Rather, a system with a negative temperature is hotter than *any* system with a positive temperature in the sense that if a negative-temperature system and a positive-temperature system come in contact, heat will flow from the negative- to the positive-temperature system.

Most familiar systems cannot achieve negative temperatures because adding energy always increases their entropy. However, some systems have a maximum amount of energy that they can hold, and as they approach that maximum energy their entropy actually begins to decrease. Because temperature is defined by the relationship between energy and entropy, such a system's temperature becomes negative, even though energy is being added. As a result, the Boltzmann factor for states of systems at negative temperature increases rather than decreases with increasing state energy. Therefore, no complete system, i.e. including the electromagnetic modes, can have negative temperatures, since there is no highest energy state, so that the sum of the probabilities of the states would diverge for negative temperatures. However, for quasi-equilibrium systems (e.g. spins out of equilibrium with the electromagnetic field) this argument does not apply, and negative effective temperatures are attainable.

On 3 January 2013, physicists announced that they had created a quantum gas made up of potassium atoms with a negative temperature in motional degrees of freedom for the first time.

History

Robert Boyle pioneered the idea of an absolute zero.

One of the first to discuss the possibility of an absolute minimal temperature was Robert Boyle. His 1665 *New Experiments and Observations touching Cold*, articulated the dispute known as the *primum frigidum*. The concept was well known among naturalists of the time. Some contended an absolute minimum temperature occurred within earth (as one of the four classical elements), others within water, others air, and some more recently within nitre. But all of them seemed to agree that, "There is some body or other that is of its own nature supremely cold and by participation of which all other bodies obtain that quality."

Limit to the "Degree of Cold"

The question whether there is a limit to the degree of coldness possible, and, if so, where the zero must be placed, was first addressed by the French physicist Guillaume Amontons in 1702, in connection with his improvements in the air-thermometer. In his instrument, temperatures were indicated by the height at which a column of mercury was sustained by a certain mass of air, the volume, or "spring", of which varied with the heat to which it was exposed. Amontons therefore argued that the zero of his thermometer would be that temperature at which the spring of the air in it was reduced to nothing. On the scale he used, the boiling-point of water was marked at +73 and the melting-point of ice at 51, so that the zero of his scale was equivalent to about −240 on the Celsius scale.

This close approximation to the modern value of −273.15 °C for the zero of the air-thermometer was further improved upon in 1779 by Johann Heinrich Lambert, who observed that −270 °C might be regarded as absolute cold.

Values of this order for the absolute zero were not, however, universally accepted about this period. Pierre-Simon Laplace and Antoine Lavoisier, in their 1780 treatise on heat, arrived at values ranging from 1,500 to 3,000 below the freezing-point of water, and thought that in any case it must be at least 600 below. John Dalton in his *Chemical Philosophy* gave ten calculations of this value, and finally adopted −3000 °C as the natural zero of temperature.

Lord Kelvin's Work

After James Prescott Joule had determined the mechanical equivalent of heat, Lord Kelvin approached the question from an entirely different point of view, and in 1848 devised a scale of absolute temperature which was independent of the properties of any particular substance and was based on Carnot's theory of the Motive Power of Heat and data published by Henri Victor Regnault. It followed from the principles on which this scale was constructed that its zero was placed at −273 °C, at almost precisely the same point as the zero of the air-thermometer. This value was not immediately accepted; values ranging from -271.1 °C to -274.5 °C, derived from laboratory measurements and observations of astronomical refraction, continued to be used in the early 20th century.

Very Low Temperatures

The rapid expansion of gases leaving the Boomerang Nebula, a bi-polar, filamentary, likely proto-planetary nebula in Centaurus, causes the lowest observed temperature outside a laboratory: 1 K

The average temperature of the universe today is approximately 2.73 kelvins (−270.42 °C; −454.76 °F), based on measurements of cosmic microwave background radiation.

Absolute zero cannot be achieved, although it is possible to reach temperatures close to it through the use of cryocoolers, dilution refrigerators, and nuclear adiabatic demagnetization. The use of laser cooling has produced temperatures less than a billionth of a kelvin. At very low temperatures in the vicinity of absolute zero, matter exhibits many unusual properties, including superconductivity, superfluidity, and Bose–Einstein condensation. To study such phenomena, scientists have worked to obtain even lower temperatures.

- The current world record was set in 1999 at 100 picokelvins (pK), or 0.000 000 000 1 of a kelvin, by cooling the nuclear spins in a piece of rhodium metal.
- In November 2000, nuclear spin temperatures below 100 pK were reported for an experiment at the Helsinki University of Technology's Low Temperature Lab in Espoo, Finland. However, this was the temperature of one particular degree of freedom – a quantum property called nuclear spin – not the overall average thermodynamic temperature for all possible degrees in freedom.
- In February 2003, the Boomerang Nebula was observed to have been releasing gases at a speed of 500,000 km/h (over 300,000 mph) for the last 1,500 years. This has cooled it down to approximately 1 K, as deduced by astronomical observation, which is the lowest natural temperature ever recorded.
- In May 2005, the European Space Agency proposed research in space to achieve femto-kelvin temperatures.
- In May 2006, the Institute of Quantum Optics at the University of Hannover gave details of technologies and benefits of femto-kelvin research in space.
- In January 2013, physicist Ulrich Schneider of the University of Munich in Germany reported to have achieved temperatures below absolute zero ("negative temperatures") in gases; the gas reportedly became hotter rather than colder.
- In September 2014, scientists in the CUORE collaboration at the Laboratori Nazionali del Gran Sasso in Italy cooled a copper vessel with a volume of one cubic meter to 0.006 kelvins (−273.144 °C; −459.659 °F) for 15 days, setting a record for the lowest temperature in the known universe over such a large contiguous volume
- In June 2015, experimental physicists at Massachusetts Institute of Technology (MIT) have successfully cooled molecules in a gas of sodium potassium to a temperature of 500 nanokelvins, and it is expected to exhibit an exotic state of matter by cooling these molecules a bit further.

Targeted Temperature Management

Targeted temperature management (TTM) previously known as therapeutic hypothermia or protective hypothermia is active treatment that tries to achieve and maintain a specific body temperature in a person for a specific duration of time in an effort to improve health outcomes during recovery after a period of stopped blood flow to the brain. This is done in an attempt to reduce the risk of tissue injury following lack of blood flow. Periods of poor blood flow may be due to cardiac arrest or the blockage of an artery by a clot as in the case of a stroke.

Targeted temperature management improves survival and brain function following resuscitation from cardiac arrest. Evidence supports its use following certain types of cardiac arrest in which an individual does not regain consciousness. Both 33 °C (91 °F) and 36 °C (97 °F) appear to result in similar outcomes. Targeted temperature management following traumatic brain injury has shown mixed results with some studies showing benefits in survival and brain function while others show no clear benefit. While associated with some complications, these are generally mild.

Targeted temperature management is thought to prevent brain injury by several methods including decreasing the brain's oxygen demand, reducing the production of neurotransmitters like glutamate, as well as reducing free radicals that might damage the brain. The lowering of body temperature may be accomplished by many means including the use of cooling blankets, cooling helmets, cooling catheters, ice packs and ice water lavage.

Medical Uses

Targeted temperature management maybe used in the following conditions:

Cardiac Arrest

The ILCOR and American Heart Association guidelines support the use of cooling following resuscitation from cardiac arrest. These recommendations were largely based on two trials from 2002 which showed improved survival and brain function when cooled between 32 °C (90 °F) to 34 °C (93 °F) after cardiac arrest. A large trial from 2013 found that an actively cooled temperature of 36 °C (97 °F) results in the same outcomes. This conclusion was supported by a 2015 meta analysis. A 2016 review found a lack of evidence to support a difference in long term quality of life.

Another trial that looked at earlier versus later cooling found no difference in survival. This trial compared cooling by cold saline in the ambulance followed by in-hospital cooling versus in-hospital cooling alone. A registry database found poor neurological outcome increased by 8% with each 5 min delay in initiating TTM and by 17% for every 30 min delay in time to target temperature.

Neonatal Encephalopathy

Hypothermia therapy for neonatal encephalopathy has been proven to improve outcomes for newborn infants affected by perinatal hypoxia-ischemia, hypoxic ischemic encephalopathy or birth asphyxia.A 2013 Cochrane review found that it is useful in full term babies with encephalopathy. Whole body or selective head cooling to 33–34 °C (91–93 °F), begun within six hours of birth and continued for 72 hours reduces mortality and reduces cerebral palsy and neurological deficits in survivors.

Adverse Effects

Possible complications may include: infection, bleeding, dysrhythmias and high blood sugar. One review found an increased risk of pneumonia and sepsis but not the overall risk of infection. Another review found a trend towards increased bleeding but no increase in severe bleeding. Hypothermia induces a "cold diuresis" which can lead to electrolyte abnormalities - specifically hypokalemia, hypomagnesaemia, and hypophosphatemia, as well as hypovolemia.

Mechanism

The earliest rationale for the effects of hypothermia as a neuroprotectant focused on the slowing of cellular metabolism resulting from a drop in body temperature. For every one degree Celsius drop in body temperature, cellular metabolism slows by 5-7%. Accordingly, most early hypotheses suggested that hypothermia reduces the harmful effects of ischemia by decreasing the body's need for oxygen. The initial emphasis on cellular metabolism explains why the early studies almost exclusively focused on the application of deep hypothermia, as these researchers believed that the therapeutic effects of hypothermia correlated directly with the extent of temperature decline.

More recent data suggests that even a modest reduction in temperature can function as a neuroprotectant, suggesting the possibility that hypothermia affects pathways that extend beyond a decrease in cellular metabolism. One plausible hypothesis centers around the series of reactions that occur following oxygen deprivation, particularly those concerning ion homeostasis. In the special case of infants suffering perinatal asphyxia it appears that apoptosis is a prominent cause of cell death and that hypothermia therapy for neonatal encephalopathy interrupts the apoptotic pathway. In general, cell death is not directly caused by oxygen deprivation, but occurs indirectly as a result of the cascade of subsequent events. Cells need oxygen to create ATP, a molecule used by cells to store energy, and cells need ATP to regulate intracellular ion levels. ATP is used to fuel both the importation of ions necessary for cellular function and the removal of ions that are harmful to cellular function. Without oxygen, cells cannot manufacture the necessary ATP to regulate ion levels and thus cannot prevent the intracellular environment from approaching the ion concentration of the outside environment. It is not oxygen deprivation itself that precipitates cell death, but rather without oxygen the cell can not make the ATP it needs to regulate ion concentrations and maintain homeostasis.

Notably, even a small drop in temperature encourages cell membrane stability during periods of oxygen deprivation. For this reason, a drop in body temperature helps prevent an influx of unwanted ions during an ischemic insult. By making the cell membrane more impermeable, hypothermia helps prevent the cascade of reactions set off by oxygen deprivation. Even moderate dips in temperature strengthen the cellular membrane, helping to minimize any disruption to the cellular environment. It is by moderating the disruption of homeostasis caused by a blockage of blood flow that many now postulate, results in hypothermia's ability to minimize the trauma resultant from ischemic injuries.

Targeted temperature management may also help to reduce reperfusion injury, damage caused by oxidative stress when the blood supply is restored to a tissue after a period of ischemia. Various inflammatory immune responses occur during reperfusion. These inflammatory responses cause increased intracranial pressure, which leads to cell injury and in some situations, cell death. Hypothermia has been shown to help moderate intracranial pressure and therefore to minimize the harmful effects of a patient's inflammatory immune responses during reperfusion. The oxidation that occurs during reperfusion also increases free radical production. Since hypothermia reduces both intracranial pressure and free radical production, this might be yet another mechanism of action for hypothermia's therapeutic effect.

Methods

There are a number of methods through which hypothermia is induced. These include: cooling catheters, cooling blankets, and application of ice applied around the body among others. As of 2013 it is unclear if one method is any better than the others. While cool intravenous fluid may be given to start the process further methods are required to keep the person cold.

Core body temperature must be measured (either via the esophagus, rectum, bladder in those who are producing urine, or within the pulmonary artery) to guide cooling. A temperature below 30 °C (86 °F) should be avoided, as adverse events increase significantly. The person should be kept at the goal temperature plus or minus half a degree Celsius for 24 hours. Rewarming should be done slowly with suggested speeds between 0.1 to 0.5 °C (0.18 to 0.90 °F) per hour.

Targeted temperature management should be started as soon as possible. The goal temperature should be reached before 8 hours. Targeted temperature management remains partially effective even when initiated as long as 6 hours after collapse.

Prior to the induction of targeted temperature management, pharmacological agents to control shivering must be administered. When body temperature drops below a certain threshold—typically around 36 °C (97 °F)—people may begin to shiver. It appears that regardless of the technique used to induce hypothermia, people begin to shiver when temperature drops below this threshold. Drugs commonly used to prevent and treat shivering in targeted temperature management include acetaminophen, buspirone, opioids including pethidine (meperidine), dexmedetomidine, fentanyl, and/or propofol. If shivering is unable to be controlled with these drugs, patients are often placed under general anesthesia and/or are given paralytic medication like vecuronium. People should be rewarmed slowly and steadily in order to avoid harmful spikes in intracranial pressure.

Cooling Catheters

Cooling catheters are inserted into a femoral vein. Cooled saline solution is circulated through either a metal coated tube or a balloon in the catheter. The saline cools the person's whole body by lowering the temperature of a person's blood. Catheters reduce temperature at rates ranging from 1.5 to 2 °C (2.7 to 3.6 °F) per hour. Through the use of the control unit, catheters can bring body temperature to within 0.1 °C (0.18 °F) of the target level. Furthermore, catheters can raise temperature at a steady rate, which helps to avoid harmful rises in intracranial pressure. A number of studies have demonstrated that targeted temperature management via catheter is safe and effective.

Adverse events associated with this invasive technique include bleeding, infection, vascular puncture, and deep vein thrombosis (DVT). Infection caused by cooling catheters is particularly harmful, as resuscitated people are highly vulnerable to the complications associated with infections. Bleeding represents a significant danger, due to a decreased clotting threshold caused by hypothermia. The risk of deep vein thrombosis may be the most pressing medical complication.

Deep vein thrombosis can be characterized as a medical event whereby a blood clot forms in a deep vein, usually the femoral vein. This condition may become potentially fatal if the clot travels to the lungs and causes a pulmonary embolism. Another potential problem with cooling catheters is the potential to block access to the femoral vein, which is a site normally used for a variety of other medical procedures, including angiography of the venous system and the right side of the heart. However, most cooling catheters are triple lumen catheters, and the majority of people post-arrest will require central venous access. Unlike non-invasive methods which can be administered by nurses, the insertion of cooling catheters must be performed by a physician fully trained and familiar with the procedure. The time delay between identifying a person who might benefit from the procedure and the arrival of an interventional radiologist or other physician to perform the insertion may minimize some of the benefit of invasive methods' more rapid cooling.

Trans Nasal Evaporative Cooling

Trans nasal evaporative cooling is a method of inducing the hypothermia process and provides a means of continuous cooling of a person throughout the early stages of targeted temperature management and during movement throughout the hospital environment. This technique uses two cannulae, inserted into a persons nasal cavity, to deliver a spray of coolant mist that evaporates directly underneath the brain and base of the skull. As blood passes through the cooling area, it reduces the temperature throughout the rest of the body.

The method is compact enough to be used at the point of cardiac arrest, during ambulance transport, or within the hospital proper. It is intended to reduce rapidly the person's temperature to below 34 °C (93 °F) while targeting the brain as the first area of cooling. Research into the device has shown cooling rates of 2.6 °C (4.7 °F) per hour in the brain (measured through infrared tympanic measurement) and 1.6 °C (2.9 °F) per hour for core body temperature reduction.

Water Blankets

With these technologies, cold water circulates through a blanket, or torso wraparound vest and leg wraps. To lower temperature with optimal speed, 70% of a person's surface area should be covered with water blankets. The treatment represents the most well studied means of controlling body temperature. Water blankets lower a person's temperature exclusively by cooling a person's skin and accordingly require no invasive procedures.

Water blankets possess several undesirable qualities. They are susceptible to leaking, which may represent an electrical hazard since they are operated in close proximity to electrically powered medical equipment. The Food and Drug Administration also has reported several cases of external cooling blankets causing significant burns to the skin of person. Other problems with external cooling include overshoot of temperature (20% of people will have overshoot), slower induction time versus internal cooling, increased compensatory response, decreased patient access, and discontinuation of cooling for invasive procedures such as the cardiac catheterization.

If therapy with water blankets is given along with two litres of cold intravenous saline, people can be cooled to 33 °C (91 °F) in 65 minutes. Most machines now come with core temperature probes. When inserted into the rectum, the core body temperature is monitored and feedback to the machine allows changes in the water blanket to achieve the desired set temperature. In the past some of the models of cooling machines have produced an overshoot in the target temperature and cooled people to levels below 32 °C (90 °F), resulting in increased adverse events. They have also rewarmed patients at too fast a rate, leading to spikes in intracranial pressure. Some of the new models have more software that attempt to prevent this overshoot by utilizing warmer water when the target temperature is close and preventing any overshoot. Some of the new machines now also have 3 rates of cooling and warming; a rewarming rate with one of these machines allows a patient to be rewarmed at a very slow rate of just 0.17 °C

(0.31 °F) an hour in the "automatic mode," allowing rewarming from 33 °C (91 °F) to 37 °C (99 °F) over 24 hours.

Cool Caps

There are a number of non-invasive head cooling caps and helmets designed to target cooling at the brain. Hypothermia caps are typically made of a synthetic such as neoprene, silicone, or polyurethane, and filled with a coolant agent such as ice or gel which is either frozen to a very cold temperature −25 to −30 °C (−13 to −22 °F) before application or continuously cooled by an auxiliary control unit. Their most notable uses are in preventing or reducing alopecia in chemotherapy, and for preventing cerebral palsy in babies born with hypoxic ischemic encephalopathy. In the continuously cooled iteration, coolant is cooled with the aid of a compressor and then pumped out into cooling caps. Circulation is controlled by temperature sensors in the cap and regulated by valves. If the temperature deviates or if other errors are detected, an alarm system is activated. The frozen iteration involves continuous application of caps filled with crylon gel cooled to −30 °C (−22 °F) to the scalp before, during and after intravenous chemotherapy. As the caps warm on the head, multiple cooled caps must be kept on hand and applied every 20 to 30 minutes.

History

Hypothermia has been applied therapeutically since antiquity. The Greek physician Hippocrates, the namesake of the Hippocratic Oath, advocated the packing of wounded soldiers in snow and ice. Napoleonic surgeon Baron Dominique Jean Larrey recorded that officers who were kept closer to the fire, survived less often than the minimally pampered infantrymen. In modern times the first medical article concerning hypothermia was published in 1945.This study focused on the effects of hypothermia on patients suffering from severe head injury. In the 1950s hypothermia received its first medical application, being used in intracerebal aneurysm surgery to create a bloodless field. Most of the early research focused on the applications of deep hypothermia, defined as a body temperature between 20–25 °C (68–77 °F). Such an extreme drop in body temperature brings with it a whole host of side effects, which made the use of deep hypothermia impractical in most clinical situations.

This period also saw sporadic investigation of more mild forms of hypothermia, with mild hypothermia being defined as a body temperature between 32–34 °C (90–93 °F). In the 1950s, Doctor Rosomoff demonstrated in dogs the positive effects of mild hypothermia after brain ischemia and traumatic brain injury. In the 1980s further animal studies indicated the ability of mild hypothermia to act as a general neuroprotectant following a blockage of blood flow to the brain. This animal data was supported by two landmark human studies that were published simultaneously in 2002 by the New England Journal of Medicine. Both studies, one occurring in Europe and the other in Australia, demonstrated the positive effects of mild hypothermia applied following cardiac arrest. Responding to this research, in 2003 the American Heart Association (AHA) and the International Liaison Committee on Resuscitation (ILCOR) endorsed the use of targeted temperature management following cardiac arrest. Currently, a growing percentage of hospitals around the world incorporate the AHA/ILCOR guidelines and include hypothermic therapies in their standard package of care for patients suffering from cardiac arrest. Some researchers go so far as to contend that hypothermia represents a better neuroprotectant following a blockage of blood to the brain than any known drug. Over this same period a particularly successful research effort showed that hypothermia is a highly effective treatment when applied to newborn infants following birth asphyxia. Meta-analysis of a number of large randomised controlled trials showed that hypothermia for 72 hours started within 6 hours of birth significantly increased the chance of survival without brain damage.

Research

TTM has been studied in several use scenarios where it has not been found to be helpful, or is still under investigation.

Stroke

There is currently no evidence supporting targeted temperature management use in humans and clinical trials have not been completed. Most of the data concerning hypothermia's effectiveness in treating stroke is limited to animal studies. These studies have focused primarily on ischemic stroke as opposed to hemorrhagic stroke, as hypothermia is associated with a lower clotting threshold. In these animal studies, hypothermia was represented an effective neuroprotectant. The use of hypothermia to control intracranial pressure (ICP) after an ischemic stroke was found to be both safe and practical.

Traumatic Brain or Spinal Cord Injury

Animal studies have shown the benefit of targeted temperature management in traumatic central nervous system (CNS) injuries. Clinical trials have shown mixed results with regards to the optimal temperature and delay of cooling. Achieving therapeutic temperatures of 33 °C (91 °F) is thought to prevent secondary neurological injuries after severe CNS trauma. A systematic review of randomised controlled trials in traumatic brain injury (TBI) suggests there is no evidence that hypothermia is beneficial.

Neurosurgery

As of 2015 hypothermia had shown no improvements in neurological outcomes or in mortality in neurosurgery.

References

- Bilstein, Roger E. (1996). Stages to Saturn: A Technological History of the Apollo/Saturn Launch Vehicles (NASA SP-4206) (The NASA History Series). NASA History Office. pp. 89–91. ISBN 0-7881-8186-6.
- ASM Handbook, Volume 4A, Steel Heat Treating Fundamentals and Processes. ASM International. 2013. pp. 382–386. ISBN 978-1-62708-011-8.
- Knuuttila, Tauno (2000). Nuclear Magnetism and Superconductivity in Rhodium. Espoo, Finland: Helsinki University of Technology. ISBN 951-22-5208-2. Retrieved 2008-02-11.
- Ian Jacobs (Dec 17, 2013). "Targeted temperature management following cardiac arrest An update" (PDF). ilcor.org. Retrieved 14 November 2014.
- Arrich, J; Holzer, M; Havel, C; Müllner, M; Herkner, H (15 February 2016). "Hypothermia for neuroprotection in adults after cardiopulmonary resuscitation.". The Cochrane database of systematic reviews. 2: CD004128. doi:10.1002/14651858.CD004128.pub4. PMID 26878327.
- Patel, JK; Parikh, PB (7 April 2016). "Association between therapeutic hypothermia and long-term quality of life in survivors of cardiac arrest: A systematic review.". Resuscitation. 103: 54–59. doi:10.1016/j.resuscitation.2016.03.024. PMID 27060536.

5

Hibernation

Some animals hibernate to escape extremely cold seasons, in order to endure the weather and also because the food is scarce. Some of the animals known for their hibernation are deer mice, skunks, bears and hedgehogs. Usually before entering hibernation, animals store enough food in them to survive. The major components of hibernation are discussed in this section.

Hibernation

Northern bat hibernating in Norway

Hibernation is a state of inactivity and metabolic depression in endotherms. Hibernation refers to a season of heterothermy that is characterized by low body temperature, slow breathing and heart rate, and low metabolic rate. Although traditionally reserved for "deep" hibernators such as rodents, the term has been redefined to include animals such as bears and is now applied based on active metabolic suppression rather than based on absolute body temperature decline. Many experts believe that the processes of daily torpor and hibernation form a continuum and utilize similar mechanisms. Hibernation during the summer months is known as aestivation. Some reptile species (ectotherms) are said to brumate, or undergo brumation, but any possible similarities between brumation and hibernation are not firmly established. Some insects, such as the wasp *Polistes exclamans*, hibernate by aggregating together in groups in protected places called hibernacula.

Bats hibernating in a silver mine

Often associated with low temperatures, the function of hibernation is to conserve energy during a period when sufficient food is unavailable. To achieve this energy saving, an endotherm will first decrease its metabolic rate, which then decreases body temperature. Hibernation may last several days, weeks, or months depending on the species, ambient temperature, time of year, and individual's body condition.

Before entering hibernation, animals need to store enough energy to last through the entire winter. Larger species become hyperphagic and eat a large amount of food and store the energy in fat deposits. In many small species, food caching replaces eating and becoming fat. Some species of mammals hibernate while gestating young, which are either born while the mother hibernates or shortly afterwards.

For example, the female polar bear goes into hibernation during the cold winter months to give birth to her offspring. She loses 15–27% of her pre-hibernation weight and uses stored fats for energy during times of food scarcity, or hibernation. It is evident that pregnant female polar bears significantly increase body mass prior to hibernation, and this increase is further reflected in the weight of their offspring. The fat accumulation prior to hibernation in female polar bears enables them to provide a sufficient and warm, nurturing environment for their newborns.

Hibernating Animals

Primates

While hibernation has long been studied in rodents, namely ground squirrels, no primate or tropical mammal was known to hibernate prior to the discovery that the fat-tailed dwarf lemur of Madagascar hibernates in tree holes for seven months of the year. Malagasy winter temperatures sometimes rise to over 30 °C (86 °F), so hibernation is not exclusively an adaptation to low ambient temperatures. The hibernation of this lemur is strongly dependent on the thermal behaviour of its tree hole: if the hole is poorly insulated, the lemur's body temperature fluctuates widely, passively following the ambient temperature; if well insulated, the body temperature stays fairly constant and the animal undergoes regular spells of arousal. Dausmann found that hypometabolism in hibernating animals is not necessarily coupled to a low body temperature.

Bears

Bears are able to recycle their proteins and urine, allowing them to both stop urinating for months and stop muscle atrophy.

Note that in some languages a specific term is used to describe the type of hibernation undergone by bears. For example, in French it is called "hivernation" instead of "hibernation".

Obligate Hibernators

Obligate hibernators are defined as animals that spontaneously, and annually, enter hibernation regardless of ambient temperature and access to food. Obligate hibernators include many species of ground squirrels, other rodents, mouse lemurs, the European hedgehog and other insectivores, monotremes, marsupials, and even butterflies such as the small tortoiseshell. These undergo what has been traditionally called "hibernation": the physiological state where the body temperature drops to near ambient (environmental) temperature, and heart and respiration rates slow drastically. The typical winter season for these hibernators is characterized by periods of torpor interrupted by periodic, euthermic arousals, wherein body temperatures and heart rates are restored to euthermic (more typical) levels. The cause and purpose of these arousals is still not clear.

The question of why hibernators may experience the periodic arousals (returns to high body temperature) has plagued researchers for decades, and while there is still no clear-cut explanation, there are myriad hypotheses on the topic. One favored hypothesis is that hibernators build a 'sleep debt' during hibernation, and so must occasionally warm up in order to sleep. This has been supported by evidence in the Arctic ground squirrel. Another theory states that the brief periods of high body temperature during hibernation are used by the animal to restore its available energy sources. Yet another theory states that the frequent returns to high body temperature allow mammals to initiate an immune response.

Hibernating Arctic ground squirrels may exhibit abdominal temperatures as low as -2.9 °C, maintaining sub-zero abdominal temperatures for more than three weeks at a time, although the temperatures at the head and neck remain at 0 °C or above.

Historically there was a question of whether or not bears truly hibernate, since they experience only a modest decline in body temperature (3–5 K) compared with what other hibernators undergo (32 K or more). Many researchers thought that their deep sleep was not comparable with true, deep hibernation. This theory has been refuted by recent research in captive black bears.

Black bear mother and cubs "denning"

Facultative Hibernation

Unlike obligate hibernators, facultative hibernators only enter hibernation when either cold stressed or food deprived, or both. A good example of the differences between the two types of hibernation can be seen among the prairie dogs: the white-tailed prairie dog is an obligate hibernator and the closely related black-tailed prairie dog is a facultative hibernator.

Hibernating Birds

Historically, Pliny the Elder believed swallows hibernated, and ornithologist Gilbert White pointed to anecdotal evidence in *The Natural History of Selborne* that indicated as much. Birds typically do not hibernate, instead utilizing torpor. One known exception is the common poorwill (*Phalaenoptilus nuttallii*), first documented by Edmund Jaeger.

Dormancy in Fish

Fish are ectothermic, and so, by definition, cannot hibernate because they cannot actively down-regulate their body temperature or their metabolic rate. However, they can experience decreased metabolic rates associated with colder environments and/or low oxygen availability (hypoxia) and can experience dormancy. For a couple of generations during the 20th century it was thought that basking sharks settled to the floor of the North Sea and became dormant. Research by Dr David Sims in 2003 dispelled this hypothesis, showing that the sharks actively traveled huge distances throughout the seasons, tracking the areas with the highest quantity of plankton. The epaulette sharks have been documented to be able to survive for long periods of time without oxygen, even being left high and dry, and at temperatures of up to 26 °C (79 °F). Other animals able to survive long periods with no or very little oxygen include the goldfish, the red-eared slider turtle, the wood frog, and the bar-headed goose. However, the ability to survive hypoxic or anoxic conditions is not the same, nor closely related, to endotherm hibernation.

Hibernation Induction Trigger

Hibernation induction trigger (HIT) is somewhat of a misnomer. Although research in the 1990s hinted at the ability to induce torpor in animals by injection of blood taken from a hibernating animal, further research has been unable to reproduce this phenomenon. Despite the inability to induce torpor, there are substances in hibernator blood that can lend protection to organs for possible transplant. Researchers were able to prolong the life of an isolated pig's heart with a HIT. This may have potentially important implications for organ transplant, as it could allow organs to survive for up to 18 or more hours, outside the human body. This would be a great improvement from the current 6 hours.

This supposed HIT is a mixture derived from serum, including at least one opioid-like substance. DADLE is an opioid that in some experiments has been shown to have similar functional properties.

Human Hibernation

There are many research projects currently investigating how to achieve "induced hibernation" in humans. This ability to hibernate humans would be useful for a number of reasons, such as saving the lives of seriously ill or injured people by temporarily putting them in a state of hibernation until treatment can be given.

Actual and anecdotal cases of suspected human hibernation or states similar to hibernation exist in the literature:

- Anna Bågenholm, a Swedish radiologist who in 1999 survived 80 minutes under ice in a frozen stream in Norway, the final 40 minutes in a state of cardiac arrest, and survived with no brain damage.
- Mitsutaka Uchikoshi, a Japanese man who survived the cold for 24 days in 2006 without food or water when he fell into a state similar to hibernation
- Paulie Hynek, who, at age 2, survived several hours of hypothermia-induced cardiac arrest and whose body temperature reached 64 °F (18 °C)
- John Smith, a 14-year-old boy who survived 15 minutes under ice in a frozen lake before paramedics arrived to pull him onto dry land and saved him.

Heterothermy

Heterothermy or heterothermia is a physiological term for animals that exhibit characteristics of both poikilothermy and homeothermy.

Definition

Heterothermic animals are those that can switch between poikilothermic and homeothermic strategies. These changes in strategies typically occur on a daily basis or on an annual basis. More often than not, it is used as a way to dissociate the fluctuating metabolic rates seen in some small mammals and birds (e.g. bats and hummingbirds), from those of traditional cold blooded animals. In many bat species, body temperature and metabolic rate are elevated only during activity. When at rest, these animals reduce their metabolisms drastically, which results in their body temperature dropping to that of the surrounding environment. This makes them homeothermic when active, and poikilothermic when at rest. This phenomenon has been termed 'daily torpor' and was intensively studied in the Djungarian hamster. During the hibernation season, this animal shows strongly reduced metabolism each day during the rest phase while they revert to endothermic metabolism during their active phase, leading to normal euthermic body temperatures (around 38 °C).

Larger mammals (e.g. ground squirrels) and bats show multi-day torpor bouts during hibernation (up to several weeks) in winter. During these multi-day torpor bouts, body temperature drops to ~1 °C above ambient temperature and metabolism may drop to about 1% of the normal endothemic metabolic rate. Even in these deep hibernators, the long periods of torpor is interrupted by bouts of endothermic metabolism, called arousals (typically lasting between 4–20 hours). These metabolic arousals cause body temperature to return to euthermic levels 35-37 °C. Most of the energy spent during hibernation is spent in arousals (70-80%), but their function remains unresolved.

Shallow hibernation patterns without arousals have been described in large mammals (like the black bear,) or under special environmental circumstances.

Regional Heterothermy

Regional heterothermy describes organisms that are able to maintain different temperature "zones" in different regions of the body. This usually occurs in the limbs, and is made possible through the use of counter-current heat exchangers, such as the rete mirabile found in tuna and certain birds. These exchangers equalize the temperature between hot arterial blood going out to the extremities and cold venous blood coming back, thus reducing heat loss. Penguins and many arctic birds use these exchangers to keep their feet at roughly the same temperature as the surrounding ice. This keeps the birds from getting stuck on an ice sheet. Other animals, like the Leatherback Sea Turtle, use the heat exchangers to gather, and retain heat generated by their muscular flippers. There are even some insects which possess this mechanism, the best-known example being bumblebees, which exhibit counter-current heat exchange at the point of constriction between the mesosoma ("thorax") and metasoma ("abdomen"); heat is retained in the thorax and lost from the abdomen. Using a very similar mecha-nism, the internal temperature of a honeybee's thorax can exceed 45 °C while in flight.

Hibernaculum (zoology)

A hibernaculum *plural form: hibernacula* (Latin, "tent for winter quarters") is a place of abode in which a creature seeks refuge, such as a bear using a cave to overwinter. Insects may hibernate to survive the winter. The word can be used to describe a variety of shelters used by various kinds of animals, for instance, bats, marmots and snakes.

Bear Gulch Cave provides a home to a colony of Townsend's big-eared bats

An artificial hibernaculum or 'Bug Hotel'

References

- "Eleva boy's story part of national tour to honor Mayo Clinics 150 years". Mayo Clinic. Archived from the original on May 11, 2015.
- Dovey, Dana (February 6, 2015). "Suspended Animation? How A Boy Survived 15 Minutes Trapped Under Ice In Frozen Lake". Medical Daily.
- Douglas Fox (March 8, 2003). "Breathless: A shark with an amazing party trick is teaching doctors how to protect the brains of stroke patients". New Scientist. Vol. 177 no. 2385. p. 46. Archived from the original on February 29, 2012. Retrieved November 9, 2006.

6

Understanding Hypothermia

Hypothermia is a condition of the muscles and tissues of the living body that has been exposed to very low levels of temperature. Hypothermia is often lethal and persons may lose their external limbs. Other conditions caused by extreme cold temperature are also discussed in this chapter such as chilblains, frostbite and trench foot.

Hypothermia

Hypothermia is defined as a body core temperature below 35.0 °C (95.0 °F). Symptoms depend on the temperature. In mild hypothermia there is shivering and mental confusion. In moderate hypothermia shivering stops and confusion increases. In severe hypothermia there may be paradoxical undressing, in which a person removes his or her clothing, as well as an increased risk of the heart stopping.

Hypothermia has two main types of causes. It classically occurs from exposure to extreme cold. It may also occur from any condition that decreases heat production or increases heat loss. Commonly this includes alcohol intoxication but may also include low blood sugar, anorexia, and advanced age, among others. Body temperature is usually maintained near a constant level of 36.5–37.5 °C (97.7–99.5 °F) through thermoregulation. Efforts to increase body temperature involve shivering, increased voluntary activity, and putting on warmer clothing. Hypothermia may be diagnosed based on either a person's symptoms in the presence of risk factors or by measuring a person's core temperature.

The treatment of mild hypothermia involves warm drinks, warm clothing and physical activity. In those with moderate hypothermia heating blankets and warmed intravenous fluids are recommended. People with moderate or severe hypothermia should be moved gently. In severe hypothermia extracorporeal membrane oxygenation (ECMO) or cardiopulmonary bypass may be useful. In those without a pulse cardiopulmonary resuscitation (CPR) is indicated along with the above measures. Rewarming is typically continued until a person's temperature is greater than 32 °C (90 °F). If there is no improvement at this point or the blood potassium level is greater than 12 mmol/liter at any time resuscitation may be discontinued.

Hypothermia is the cause of at least 1500 deaths a year in the United States. It is more common in older people and males. One of the lowest documented body temperatures from which someone with accidental hypothermia has survived is 13.0 °C (55.4 °F) in a near-drowning of a 7-year-old girl in Sweden. Survival after more than six hours of CPR has been described. In those in whom ECMO or bypass is used survival is around 50%. Deaths due to hypothermia have played an important role in many wars. Hyperthermia is the opposite of hypothermia, being an increased body temperature due to failed thermoregulation.

Classification

Hypothermia is often defined as any body temperature below 35.0 °C (95.0 °F). With this method it is divided into degrees of severity based on the core temperature.

Another classification system, the Swiss staging system, divides hypothermia based on the presenting symptoms which is preferred when it is not possible to determine an accurate core temperature.

Other cold-related injuries that can be present either alone or in combination with hypothermia include:

- Chilblains, superficial ulcers of the skin that occur when a predisposed individual is repeatedly exposed to cold.
- Frostbite, the freezing and destruction of tissue.
- Frostnip, a superficial cooling of tissues without cellular destruction.
- Trench foot or immersion foot, caused by repetitive exposure to water at non-freezing temperatures.

The normal human body temperature is often stated as 36.5–37.5 °C (97.7–99.5 °F). Hyperthermia and fever, are defined as a temperature of greater than 37.5–38.3 °C (99.5–100.9 °F).

Signs and Symptoms

Signs and symptoms vary depending on the degree of hypothermia, and may be divided by the three stages of severity. Infants with hypothermia may feel cold when touched, with bright red skin and unusual lack of energy.

Mild

Symptoms of mild hypothermia may be vague, with sympathetic nervous system excitation (shivering, high blood pressure, fast heart rate, fast respiratory rate, and contraction of blood vessels). These are all physiological responses to preserve heat. Increased urine production due to cold, mental confusion, and hepatic dysfunction may also be present. Hyperglycemia may be present, as glucose consumption by cells and insulin secretion both decrease, and tissue sensitivity to insulin may be blunted. Sympathetic activation also releases glucose from the liver. In many cases, however, especially in alcoholic patients, hypoglycemia appears to be a more common presentation. Hypoglycemia is also found in many hypothermic patients, because hypothermia may be a result of hypoglycemia.

Moderate

Low body temperature results in shivering becoming more violent. Muscle mis-coordination becomes apparent. Movements are slow and labored, accompanied by a stumbling pace and mild confusion, although the person may appear alert. Surface blood vessels contract further as the body focuses its remaining resources on keeping the vital organs warm. The subject becomes pale. Lips, ears, fingers, and toes may become blue.

Severe

As the temperature decreases, further physiological systems falter and heart rate, respiratory rate, and blood pressure all decrease. This results in an expected heart rate in the 30s at a temperature of 28 °C (82 °F).

Difficulty speaking, sluggish thinking, and amnesia s tart to appear; inability to use hands and stumbling are also usually present. Cellular metabolic processes shut down. Below 30 °C (86 °F), the exposed skin becomes blue and puffy, muscle coordination very poor, and walking almost impossible, and the person exhibits incoherent/irrational behavior, including terminal burrowing or even stupor. Pulse and res-piration rates decrease significantly, but fast heart rates (ventricular tachycardia, atrial fibrillation) can also occur. Atrial fibrillation is not typically a concern in and of itself. Major organs fail. Clinical death occurs.

Paradoxical Undressing

Twenty to fifty percent of hypothermia deaths are associated with paradoxical undressing. This typically occurs during moderate to severe hypothermia, as the person becomes disoriented, confused, and combative. They may begin discarding their clothing, which, in turn, increases the rate of heat loss.

Rescuers who are trained in mountain survival techniques are taught to expect this; however, some may assume incorrectly that urban victims of hypothermia have been subjected to a sexual assault.

One explanation for the effect is a cold-induced malfunction of the hypothalamus, the part of the brain that regulates body temperature. Another explanation is that the muscles contracting peripheral blood vessels become exhausted (known as a loss of vasomotor tone) and relax, leading to a sudden surge of blood (and heat) to the extremities, fooling the person into feeling overheated.

Terminal Burrowing

An apparent self-protective behaviour known as terminal burrowing, or hide-and-die syndrome, occurs in the final stages of hypothermia. The afflicted will enter small, enclosed spaces, such as underneath beds or behind wardrobes. It is often associated with paradoxical undressing. Researchers in Germany claim this is "obviously an autonomous process of the brain stem, which is triggered in the final state of hypothermia and produces a primitive and burrowing-like behavior of protection, as seen in hibernating animals." This happens mostly in cases where temperature drops slowly.

Causes

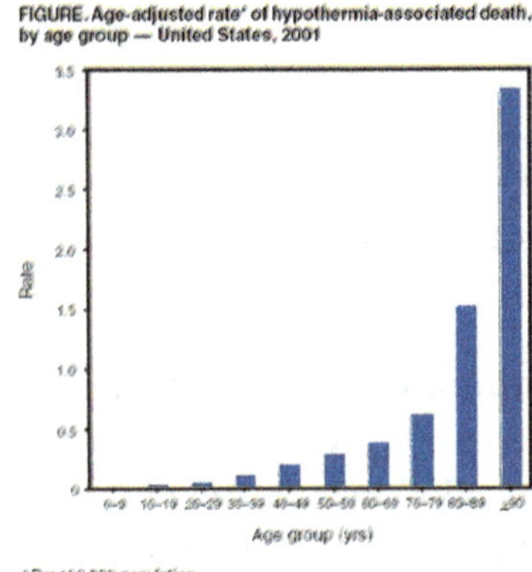

The rate of hypothermia is strongly related to age in the United States.

Hypothermia usually occurs from exposure to low temperatures, and is frequently complicated by alcohol consumption. Any condition that decreases heat production, increases heat loss, or impairs thermoregulation, however, may contribute. Thus, hypothermia risk factors include: substance abuse (including alcohol abuse), homelessness, any condition that affects judgment (such as hypoglycemia), the extremes of age, poor clothing, chronic medical conditions (such as hypothyroidism and sepsis), and living in a cold environment. Hypothermia occurs frequently in major trauma, and is also observed in severe cases of anorexia nervosa.

Alcohol

Alcohol consumption increases the risk of hypothermia by its action as a vasodilator. It increases blood flow to the skin and extremities, making a person *feel* warm, while increasing heat loss. Between 33% and 73% of hypothermia cases are complicated by alcohol.

Poverty

In the UK, 28,354 cases of hypothermia were treated in 2012/13 – an increase of 25%

from the previous year. Some cases of hypothermia death, as well as other preventable deaths, happen because poor people cannot easily afford to keep warm. Rising fuel bills have increased the numbers who have difficulty paying for adequate heating in the UK. Some pensioners and disabled people are at risk because they do not work and cannot easily get out of their homes. Better heat insulation can help.

Water Immersion

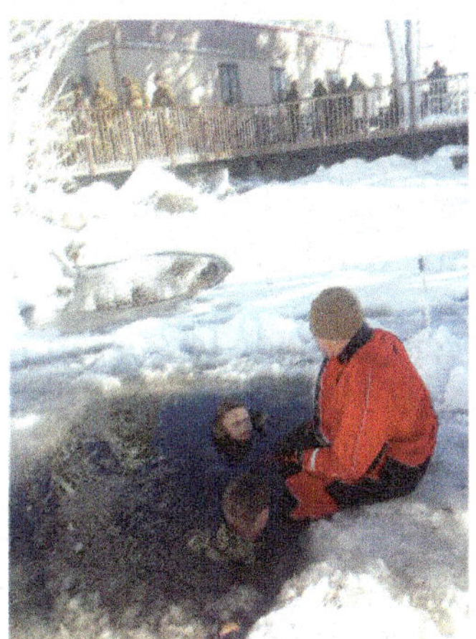

Two American marines participating in an immersion hypothermia exercise

Hypothermia continues to be a major limitation to swimming or diving in cold water. The reduction in finger dexterity due to pain or numbness decreases general safety and work capacity, which consequently increases the risk of other injuries.

Other factors predisposing to immersion hypothermia include dehydration, inadequate rewarming between repetitive dives, starting a dive while wearing cold, wet dry suit undergarments, sweating with work, inadequate thermal insulation (for example, thin dry suit undergarment), and poor physical conditioning.

Heat is lost much more quickly in water than in air. Thus, water temperatures that would be quite reasonable as outdoor air temperatures can lead to hypothermia in survivors, although this is not usually the direct clinical cause of death for those who are not rescued. A water temperature of 10 °C (50 °F) can lead to death in as little as one hour, and water temperatures near freezing can cause death in as little as 15 minutes. A notable example of this occurred during the sinking of the *Titanic*, when most people who entered the −2 °C (28 °F) water died in 15–30 minutes.

The actual cause of death in cold water is usually the bodily reactions to heat loss and to freezing water, rather than hypothermia (loss of core temperature) itself. For example, plunged into freezing seas, around 20% of victims die within 2 minutes from cold shock (uncontrolled rapid breathing, and gasping, causing water inhalation, massive increase in blood pressure and cardiac strain leading to cardiac arrest, and panic); another 50% die within 15–30 minutes from cold incapacitation (inability to use or control limbs and hands for swimming or gripping, as the body "protectively" shuts down the peripheral muscles of the limbs to protect its core). Exhaustion and unconsciousness cause drowning, claiming the rest within a similar time.

Pathophysiology

Heat is primarily generated in muscle tissue, including the heart, and in the liver, while it is lost through the skin (90%) and lungs (10%). Heat production may be increased 2 to 4 fold through muscle contractions (i.e. exercise and shivering). The rate of heat loss is determined, as with any object, by convection, conduction, and radiation. The rates of these can be affected by body mass index, body surface area to volume ratios, clothing and other environmental conditions.

Many changes to physiology occur as body temperature decreases. These occur in the cardiovascular system leading to the Osborn J wave and other dysrhythmias, decreased CNS electrical activity, cold diuresis, and non-cardiogenic pulmonary edema.

Research has shown that glomerular filtration rate (GFR) decreases as a result of hypothermia. In essence, Hypothermia increases preglomerular vasoconstriction, thus decreasing both renal blood flow (RBF) and GFR.

Diagnosis

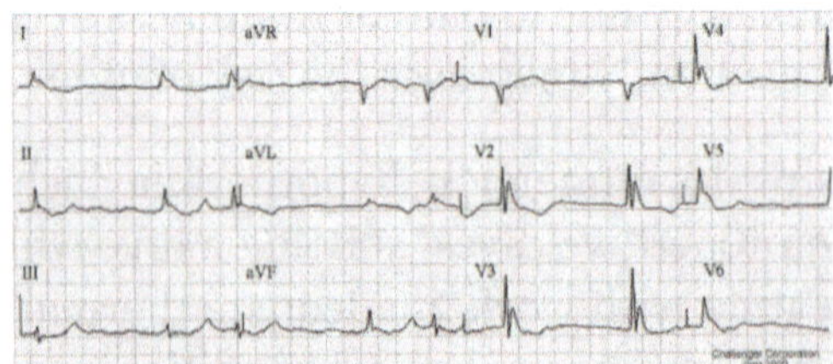

Atrial fibrillation and Osborn J waves in a person with hypothermia. Note what could be mistaken for ST elevation.

Accurate determination of core temperature often requires a special low temperature thermometer, as most clinical thermometers do not measure accurately below 34.4 °C (93.9 °F). A low temperature thermometer can be placed in the rectum, esophagus or the bladder. Esophageal measurements are the most accurate and are recommended once a person is intubated. Other methods of measurement such as in the mouth, under the arm, or using an infrared ear thermometer are often not accurate.

As a hypothermic person's heart rate may be very slow, prolonged feeling for a pulse could be required before detecting. In 2005, the American Heart Association recommended at least 30–45 seconds to verify the absence of a pulse before initiating CPR. Others recommend a 60-second check.

The classical ECG finding of hypothermia is the Osborn J wave. Also, ventricular fibrillation frequently occurs below 28 °C (82 °F) and asystole below 20 °C (68 °F). The Osborn J may look very similar to those of an acute ST elevation myocardial infarction. Thrombolysis as a reaction to the presence of Osborn J waves is not indicated, as it would only worsen the underlying coagulopathy caused by hypothermia.

Prevention

Appropriate clothing helps to prevent hypothermia. Synthetic and wool fabrics are superior to cotton as they provide better insulation when wet and dry. Some synthetic fabrics, such as polypropylene and polyester, are used in clothing designed to wick perspiration away from the body, such as liner socks and moisture-wicking undergarments. Clothing should be loose fitting, as tight clothing reduces the circulation of warm blood. In planning outdoor activity, prepare appropriately for possible cold weather. Those who drink alcohol before or during outdoor activity should ensure at least one sober person is present responsible for safety.

Covering the head is effective, but no more effective than covering any other part of the body. While common folklore says that people lose most of their heat through their heads, heat loss from the head is no more significant than that from other uncovered parts of the body. However, heat loss from the head is significant in infants, whose head is larger relative to the rest of the body than in adults. Several studies have shown that for uncovered infants, lined hats significantly reduce heat loss and thermal stress. Children have a larger surface area per unit mass, and other things being equal should have one more layer of clothing than adults in similar conditions, and the time they spend in cold environments should be limited. However children are often more active than adults, and may generate more heat. In both adults and children, overexertion causes sweating and thus increases heat loss.

Building a shelter can aid survival where there is danger of death from exposure. Shelters can be of many different types, metal can conduct heat away from the occupants and is sometimes best avoided. The shelter should not be too big so body warmth stays near the occupants. Good ventilation is essential especially if a fire will be lit in the shelter. Fires should be put out before the occupants sleep to prevent carbon monoxide poisoning. People caught in very cold, snowy conditions can build an igloo or snow cave to shelter.

The United States Coast Guard promotes using life vests to protect against hypothermia through the 50/50/50 rule: If someone is in 50 °F (10 °C) water for 50 minutes, he/she has a 50 percent better chance of survival if wearing a life jacket. A heat escape lessening position can be used to increase survival in cold water.

Babies should sleep at 16-20 °C (61-68 °F) and housebound people should be checked regularly to make sure the temperature of the home is sufficient.

Management

Degree	Rewarming technique
Mild (stage 1)	Passive rewarming
Moderate (stage 2)	Active external rewarming
Severe (stage 3 and 4)	Active internal rewarming

Aggressiveness of treatment is matched to the degree of hypothermia. Treatment ranges from noninvasive, passive external warming to active external rewarming, to active core rewarming. In severe cases resuscitation begins with simultaneous removal from the cold environment and management of the airway, breathing, and circulation. Rapid rewarming is then commenced. Moving the person as little and as gently as possible is recommended as aggressive handling may increase risks of a dysrhythmia.

Hypoglycemia is a frequent complication and needs to be tested for and treated. Intravenous thiamine and glucose is often recommended, as many causes of hypothermia are complicated by Wernicke's encephalopathy.

The UK National Health Service advises the lay public against putting a person in a hot bath, massaging their arms and legs, using a heating pad, or giving them alcohol. These measures can cause blood to be directed to the skin, causing a fall in blood pressure to vital organs, potentially resulting in death.

Rewarming

Rewarming can be done with a number of methods including passive external rewarming, active external rewarming, and active internal rewarming. Passive external rewarming involves the use of a person's own ability to generate heat by providing properly insulated dry clothing and moving to a warm environment. It is recommended for those with mild hypothermia.

Active external rewarming involves applying warming devices externally, such as a heating blanket. These may function by warmed forced air (Bair Hugger is a commonly used device), chemical reactions, or electricity. In wilderness environments, hypothermia may be helped by placing a hot water bottle in both armpits and groin. These methods are recommended for moderate hypothermia. Active core rewarming involves the use of intravenous warmed fluids, irrigation of body cavities with warmed fluids (the chest or abdomen), use of warm humidified inhaled air, or use of extracorporeal rewarming such as via a heart lung machine or extracorporeal membrane oxygenation (ECMO). Extracorporeal rewarming is the fastest method for those with severe hypothermia. Survival rates with normal mental functioning have been reported at around 50%. Chest irrigation is recommended if bypass or ECMO is not possible.

Rewarming shock (or rewarming collapse) is a sudden drop in blood pressure in combination with a low cardiac output which may occur during active treatment of a severely hypothermic person. There was a theoretical concern that external rewarming rather than internal rewarming may increase the risk. These concerns were partly believed to be due to afterdrop, a situation detected during laboratory experiments where there is a continued decrease in core temperature after rewarming has been started. Recent studies have not supported these concerns, and problems are not found with active external rewarming.

Fluids

Warm sweetened liquids can be given provided the person is alert and can swallow. Many recommend that alcohol and drinks with lots of caffeine be avoided. As most people are moderately dehydrated due to cold-induced diuresis, warmed intravenous fluids to a temperature of 38–45 °C (100–113 °F) are often recommended.

Cardiac Arrest

In those without signs of life cardiopulmonary resuscitation (CPR) should be continued during active rewarming. For ventricular fibrillation or ventricular tachycardia, a single defibrillation should be attempted. People with severe hypothermia however may not respond to pacing or defibrillation. It is not known if further defibrillation should be withheld until the core temperature reaches 30 °C (86 °F). In Europe epinephrine is not recommended until the temperature reaches 30 °C (86 °F) while the American Heart Association recommended up to three doses of epinephrine before 30 °C (86 °F) is reached. Once a temperature of 30 °C (86 °F) is reached, normal ACLS protocols should be followed.

Prognosis

It is usually recommended not to declare a person dead until their body is warmed to a near normal body temperature of greater than 32 °C (90 °F), since extreme hypothermia can suppress heart and brain function. Exceptions include if there is an obvious fatal injuries or the chest is frozen so that it cannot be compressed. If a person was buried in an avalanche for more than 35 minutes and is found with a mouth packed full of snow without a pulse stopping early may also be reasonable. This is also the case if a person's blood potassium is greater than 12 mmol/l.

Those who are stiff with pupils that do not move may survive if treated aggressively. Survival with good function also occasionally occurs even after the need for hours of CPR. Children who have a near-drowning accidents in water near 0 °C (32 °F) can occasionally be revived even over an hour after losing consciousness. The cold water lowers metabolism, allowing the brain to withstand a much longer period of hypoxia. While survival is possible, mortality from severe or profound hypothermia remains high despite optimal treatment. Studies estimate mortality at between 38% and 75%.

In those who have hypothermia due to another underlying health problem, when death occurs it is frequently from that underlying health problem.

Epidemiology

In the past, hypothermia occurred most frequently in homeless people, but recreational exposure to cold environments is now the main cause of hypothermia. Between 1995 and 2004 in the United States, an average of 1560 cold-related emergency department visits occurred per year and in the years 1999 to 2004, an average of 647 people died per year due to hypothermia.

History

The armies of Napoleon retreat from Russia in 1812.

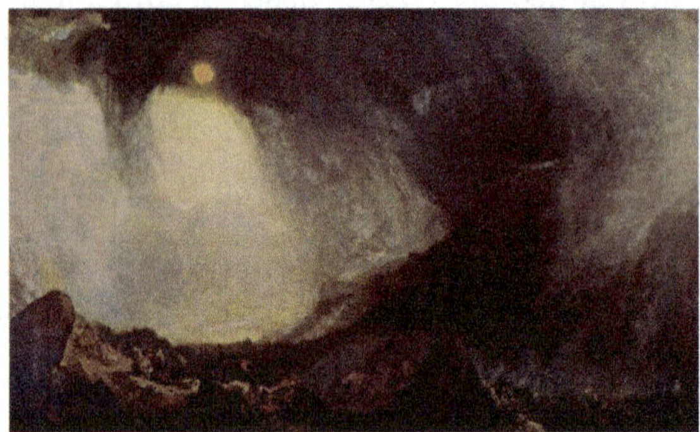

Snow-storm: Hannibal and His Army Crossing the Alps, J. M. W. Turner

Hypothermia has played a major role in the success or failure of many military campaigns, from Hannibal's loss of nearly half his men in the Second Punic War (218 B.C.) to the near destruction of Napoleon's armies in Russia in 1812. Men wandered around confused by hypothermia, some lost consciousness and died, others shivered, later developed torpor, and tended to sleep. Others too weak to walk fell on their knees; some stayed that way some time resisting death. The pulse of some was weak and hard to detect; others groaned; yet others had eyes open and wild with quiet delirium. Loss of life to hypothermia in Russian regions continued through the first and second world wars, especially in the Battle of Stalingrad.

Civilian examples of deaths caused by hypothermia occurred during the sinkings of the RMS *Titanic* and RMS *Lusitania*, and more recently of the MS *Estonia*.

Antarctic explorers developed hypothermia; Ernest Shackleton and his team measured body temperatures "below 94.2°, which spells death at home", though this probably referred to oral temperatures rather than core temperature and corresponded to mild hypothermia. One of Scott's team, Atkinson, became confused through hypothermia.

Nazi human experimentation during World War II amounting to medical torture included hypothermia experiments, which killed many victims. There were 360 to 400 experiments and 280 to 300 subjects, indicating some had more than one experiment performed on them. Various methods of rewarming were attempted, "One assistant later testified that some victims were thrown into boiling water for rewarming".

Other Animals

Many animals other than humans often induce hypothermia during hibernation or torpor.

Water bears (Tardigrade), microscopic multicellular organisms, can survive freezing at low temperatures by replacing most of their internal water with the sugar trehalose, preventing the crystallization that otherwise damages cell membranes.

Chilblains

Chilblains — also known as pernio and perniosis — is a medical condition that occurs when a predisposed individual is exposed to cold and humidity, causing tissue damage. It is often confused with frostbite and trench foot. Damage to capillary beds in the skin causes redness, itching, inflammation, and sometimes blisters. Chilblains can be reduced by keeping the feet and hands warm in cold weather, and avoiding extreme temperature changes. Chilblains can be idiopathic (spontaneous and unrelated to another disease), but may also be a manifestation of another serious medical condition that needs to be investigated. A history of chilblains is suggestive of a connective tissue disease (such as lupus). Chilblains in infants, together with severe neurologic disease and unexplained fevers, can be seen in Aicardi–Goutières syndrome, a rare inherited condition.

Signs and Symptoms

The areas most affected are the toes, fingers, earlobes, nose.

- Blistering of affected area

- Burning and itching sensation in extremities
- Dermatitis in extremities
- Digital ulceration (severe cases only)
- Erythema (blanchable redness of the skin)
- Pain in affected area
- Skin discoloration, red to dark blue

Chilblains usually heal within 7–14 days.

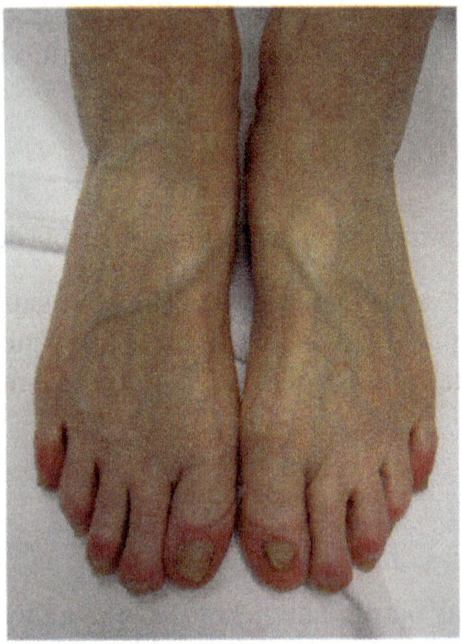

Chilblains from excessively icing the feet

Prevention

Exposure

- Avoid rapid changes in temperature (including from cold to hot).
- Wear warm shoes, socks and gloves.
- Wear a hat and a scarf to protect the ears and the nose.
- Avoid tight fitting socks/shoes.
- Place cotton wool between the toes to improve circulation.
- Recommend soaking in warm water with Epsom salts for 15–20 minutes, 3–4 times a day. Avoid very hot water.

Other

- Healthy diet, low in inflammatory foods
- Exercise at least four times a week to improve circulation

Treatment

- Keep area warm, and avoid any extreme temperature changes (including very hot water).
- Keep affected area dry.
- Use a topical steroid cream to relieve itch.
- Nifedipine, a vasodilator, may be used in more severe or recurrent cases. Vasodilation helps reduce pain, facilitate healing and prevent recurrences. It is typically available in an oral pill but can be compounded into a topical formula.
- Diltiazem, a vasodilator, may also be used.
- Apply a mixture of friar's balsam and a weak iodine solution.
- Avoid restricting the affected area.
- A common tradition of Hispanic America recommended apply a warm garlic on the chilblains.

History

The medieval *Bald's Leechbook* recommended that chilblains be treated with a mix of eggs, wine, and fennel root.

Frostbite

Frostbite or cold burn is the medical condition in which localized damage is caused to skin and other tissues due to freezing. Frostbite is most likely to happen in body parts farthest from the heart and those with large exposed areas. The initial stages of frostbite are sometimes called frostnip.

Classification

The several classifications for tissue damage caused by extreme cold include:

- Frostnip is a superficial cooling of tissues without cellular destruction.

- Chilblains are superficial ulcers of the skin that occur when a predisposed individual is repeatedly exposed to cold.
- Frostbite involves tissue destruction.

Signs and Symptoms

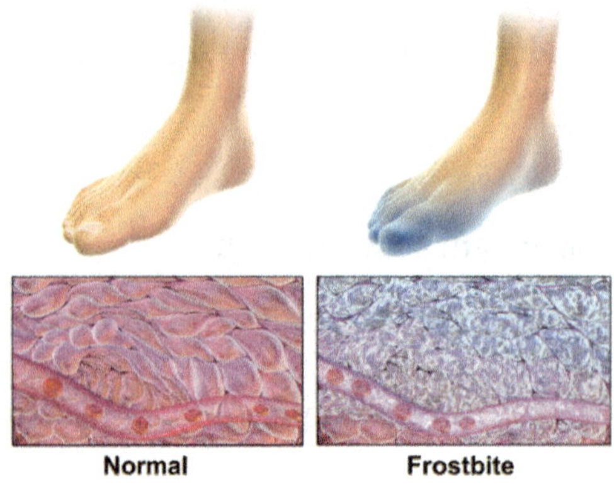

Frostbite

At or below 0 °C (32 °F), blood vessels close to the skin start to constrict, and blood is shunted away from the extremities via the action of glomus bodies. The same response may also be a result of exposure to high winds. This constriction helps to preserve core body temperature. In extreme cold, or when the body is exposed to cold for long periods, this protective strategy can reduce blood flow in some areas of the body to dangerously low levels. This lack of blood leads to the eventual freezing and death of skin tissue in the affected areas. Of the four degrees of frostbite, each has varying degrees of pain.

First Degree

This is called frostnip and only affects the surface of the skin, which is frozen. On the onset, itching and pain occur, and then the skin develops white, red, and yellow patches and becomes numb. The area affected by frostnip usually does not become permanently damaged, as only the skin's top layers are affected. Long-term insensitivity to both heat and cold can sometimes happen after suffering from frostnip.

Second Degree

If freezing continues, the skin may freeze and harden, but the deep tissues are not affected and remain soft and normal. Second-degree injury usually blisters 1–2 days after becoming frozen. The blisters may become hard and blackened, but usually appear worse than they are. Most of the injuries heal in one month, but the area may become permanently insensitive to both heat and cold.

Third and Fourth Degrees

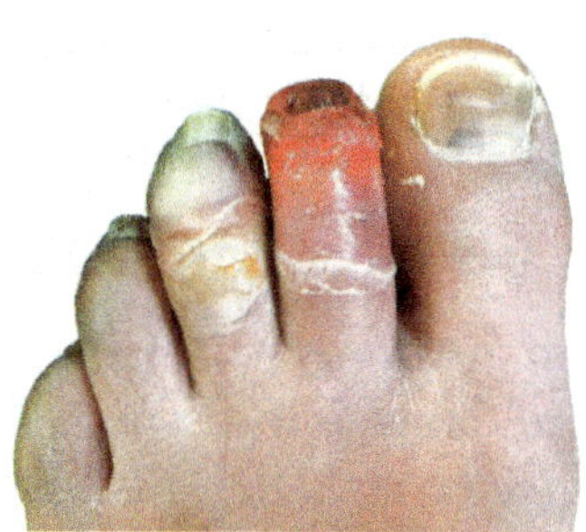

Frostbite 12 days later

If the area freezes further, deep frostbite occurs. The muscles, tendons, blood vessels, and nerves all freeze. The skin is hard, feels waxy, and use of the area is lost temporarily, and in severe cases, permanently. The deep frostbite results in areas of purplish blisters which turn black and which are generally blood-filled. Nerve damage in the area can result in a loss of feeling. This extreme frostbite may result in fingers and toes being amputated if the area becomes gangrenous. If the frostbite has gone on untreated, they may fall off. The extent of the damage done to the area by the freezing process of the frostbite may take several months to assess, and this often delays surgery to remove the dead tissue.

Causes

Inadequate blood circulation when the ambient temperature is below freezing point leads to frostbite. This can be because the body is constricting circulation to extremities on its own to preserve core temperature and fight hypothermia. In this scenario, the same factors that can lead to hypothermia (extreme cold, inadequate clothing, wet clothes, wind chill) can contribute to frostbite. Poor circulation can also be caused by other factors such as tight clothing or boots, cramped positions, fatigue, certain medications, smoking, alcohol use, or diseases that affect the blood vessels, such as diabetes.

Exposure to liquid nitrogen and other cryogenic liquids can cause frostbite, as well as prolonged contact with aerosol sprays .

Risk Factors

Risk factors for frostbite include using beta-blockers and having conditions such as diabetes and peripheral neuropathy.

Pathophysiology

Although drop in temperature and ischemia is considered to be the basic mechanism, presence of frequently observed capillary thrombi in the lesions suggests a more complicated mechanism than pure vaso-constriction. Cold temperature can cause metabolic abnormality including but not limited to crystal formation within extracellular and intracelluar fluids affecting cell function and structure including necrosis. Cold would particularly affect the platelets and enhances platelet aggregation. This mechanism seems to be particularly important during rewarming and has led to the idea of using thrombolytic medication as method of treatment.

Treatment

The decision to thaw is based on proximity to a stable, warm environment. If rewarmed tissue ends up refreezing, more damage to tissue will be done. Excessive movement of frostbitten tissue can cause ice crystals that have formed in the tissue to do further damage. Splinting or wrapping frostbitten extremities are, therefore, recommended to prevent such movement. For this reason, rubbing, massaging, shaking, or otherwise applying physical force to frostbitten tissues in an attempt to rewarm them can be harmful.

Warming can be achieved in one of two ways:

Passive rewarming involves using body heat or ambient room temperature to aid the person's body in rewarming itself. This includes wrapping in blankets or moving to a warmer environment.

Active rewarming is the direct addition of heat to a person, usually in addition to the treatments included in passive rewarming. Active rewarming requires more equipment, and therefore may be difficult to perform in the prehospital environment. When performed, active rewarming seeks to warm the injured tissue as quickly as possible without burning. This is desirable, because the faster tissue is thawed, the less tissue damage occurs. Active rewarming is usually achieved by immersing the injured tissue in a water-bath that is held between 40 and 42 °C (104 and 108 °F). Warming of peripheral tissues can increase blood flow from these areas back to the body's core. This may produce a decrease in the body's core temperature and increase the risk of abnormal heart rhythms.

Surgery

Debridement or amputation of necrotic tissue is usually delayed. This has led to the adage "Frozen in January, amputate in July", with exceptions only being made for signs of infections or gas gangrene.

Prognosis

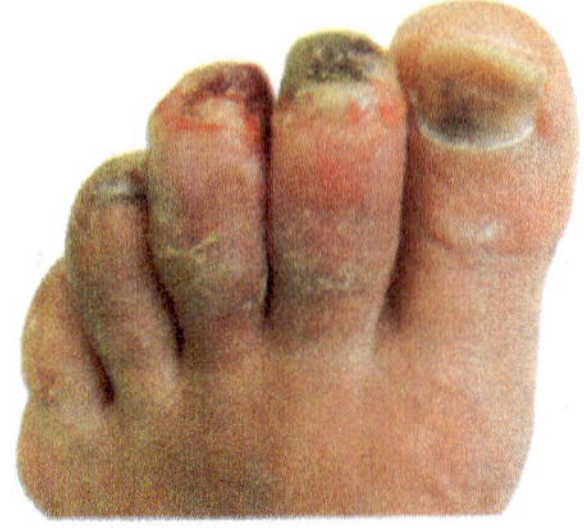

3 weeks after initial frostbite

A number of long term sequelae can occur after frostbite. These include transient or permanent changes in sensation, paresthesia, increased sweating, cancers, and bone destruction/arthritis in the area affected.

Research

Evidence is insufficient to determine whether or not hyperbaric oxygen therapy as an adjunctive treatment can assist in tissue salvage. Cases have been reported, but no randomized control trial has been performed on humans.

Medical sympathectomy using intravenous reserpine has also been attempted with limited success. Studies have suggested that administration of tissue plasminogen activator (tPa) either intravenously or intra-arterially may decrease the likelihood of eventual need for amputation.

While extreme weather conditions (cold and wind) increase the risk of frostbite, certain individuals and population groups appear more disposed to frostbite.

Trench Foot

Trench foot is a medical condition caused by prolonged exposure of the feet to damp, unsanitary, and cold conditions. It is one of many immersion foot syndromes. The use of the word *trench* in the name of this condition is a reference to trench warfare, mainly associated with World War I, which started in 1914.

Signs and Symptoms

Affected feet may become numb, affected by erythema (turning red) or cyanosis (turning blue) as a result of poor blood supply, and may begin emanating a decaying odour if the early stages of necrosis (tissue death) set in. As the condition worsens, feet may also begin to swell. Advanced trench foot often involves blisters and open sores, which lead to fungal infections; this is sometimes called tropical ulcer (jungle rot).

If left untreated, trench foot usually results in gangrene, which may require amputation. If trench foot is treated properly, complete recovery is normal, though it is marked by severe short-term pain when feeling returns.

Causes

Unlike frostbite, trench foot does not require freezing temperatures; it can occur in temperatures up to 16° Celsius (about 60° Fahrenheit) and within as few as 13 hours. The mechanism of tissue damage is not fully understood. Excessive sweating (hyperhidrosis) has long been regarded as a contributory cause; unsanitary, cold, and wet conditions can also cause trench foot.

Prevention

Trench foot can be prevented by keeping the feet clean, warm, and dry. It was also discovered in World War I that a key preventive measure was regular foot inspections; soldiers would be paired and each made responsible for the feet of the other, and they would generally apply whale oil to prevent trench foot. If left to their own devices, soldiers might neglect to take off their own boots and socks to dry their feet each day, but if it were the responsibility of another, this became less likely. Later on in the war, instances of trench foot began to decrease, probably as a result of the introduction of the aforementioned measures; of wooden duckboards to cover the muddy, wet, cold ground of the trenches; and of the increased practice of troop rotation, which kept soldiers from prolonged time at the front.

History

Trench foot was first documented by Napoleon's army in 1812. It became prevalent during the retreat from Russia and was first described by French army surgeon Dominique Jean Larrey. It was also a problem for soldiers engaged in trench warfare during the winters of World War I (hence the name),.

Trench foot made a reappearance in the British Army, during the Falklands War in 1982. The causes were the cold, wet conditions and insufficiently waterproof DMS boots.

Some people were even reported to have developed trench foot at the 1998 and 2007 Glastonbury Festivals, the 2009 and 2013 Leeds Festivals, as well as the 2012 Download Festival, as a result of the sustained cold, wet, and muddy conditions at the events.

Popular Culture

In *Hell on Wheels* season 1, episode 8 ("Derailed"), Eva warns Lily to lay down planks in her tent, to avoid getting trench foot. Lily replies that she's well acquainted with trench foot, having spent a year in the field as a surveyor for the Union Pacific Railroad.

In the HBO miniseries *Band of Brothers*, Episode 6 features an instance of trench foot during the Siege of Bastogne.

Thermoregulation

Thermoregulation is the ability of an organism to keep its body temperature within certain boundaries, even when the surrounding temperature is very different. A thermoconforming organism, by contrast, simply adopts the surrounding temperature as its own body temperature, thus avoiding the need for internal thermoregulation. The internal thermoregulation process is one aspect of homeostasis: a state of dynamic stability in an organism's internal conditions, maintained far from thermal equilibrium with its environment (the study of such processes in zoology has been called physiological ecology). If the body is unable to maintain a normal temperature and it increases significantly above normal, a condition known as hyperthermia occurs. For humans, this occurs when the body is exposed to constant temperatures of approximately 55 °C (131 °F), and with prolonged exposure (longer than a few hours) at this temperature and up to around 75 °C (167 °F) death is almost inevitable. Humans may also experience lethal hyperthermia when the wet bulb temperature is sustained above 35 °C (95 °F) for six hours. The opposite condition, when body temperature decreases below normal levels, is known as hypothermia.

It was not until the introduction of thermometers that any exact data on the temperature of animals could be obtained. It was then found that local differences were present, since heat production and heat loss vary considerably in different parts of the body, although the circulation of the blood tends to bring about a mean temperature of the internal parts. Hence it is important to identify the parts of the body that most closely reflect the temperature of the internal organs. Also, for such results to be comparable, the measurements must be conducted under comparable conditions. The rectum has traditionally been considered to reflect most accurately the temperature of internal parts, or in some cases of sex or species, the vagina, uterus or bladder.

Occasionally the temperature of the urine as it leaves the urethra may be of use in measuring body temperature. More often the temperature is taken in the mouth, axilla, ear or groin.

Some animals undergo one of various forms of dormancy where the thermoregulation process temporarily allows the body temperature to drop, thereby conserving energy. Examples include hibernating bears and torpor in bats.

Classification of Animals by Thermal Characteristics

Endothermy vs. Ectothermy

Thermoregulation in organisms runs along a spectrum from endothermy to ectothermy. Endotherms create most of their heat via metabolic processes, and are colloquially referred to as warm-blooded. Ectotherms use external sources of temperature to regulate their body temperatures. They are colloquially referred to as cold-blooded despite the fact that body temperatures often stay within the same temperature ranges as warm-blooded animals.

Ectotherms

Seeking shade is one method of cooling. Here sooty tern chicks are using a black-footed albatross chick for shade.

Ectothermic Cooling

- Vaporization:
  - Evaporation of sweat and other bodily fluids.
- Convection:
  - Increasing blood flow to body surfaces to maximize heat loss.
- Conduction:
  - Losing heat by being in contact with a colder surface. For instance:
    - Lying on cool ground.
    - Staying wet in a river, lake or sea.
    - Covering in cool mud.
- Radiation:
  - releasing heat by radiating it away from the body.

Ectothermic Heating (or Minimizing Heat Loss)

- Convection:
  - Climbing to higher ground up trees, ridges, rocks.
  - Entering a warm water or air current.
  - Building an insulated nest or burrow.
- Conduction:

- Lying on a hot surface.

- Radiation:
    - Lying in the sun (heating this way is affected by the body's angle in relation to the sun).
    - Folding skin to reduce exposure.
    - Concealing wing surfaces.
    - Exposing wing surfaces.
- Insulation:
    - Changing shape to alter surface/volume ratio.
    - Inflating the body.

Thermographic image of a snake around an arm

To cope with low temperatures, some fish have developed the ability to remain functional even when the water temperature is below freezing; some use natural antifreeze or antifreeze proteins to resist ice crystal formation in their tissues. Amphibians and reptiles cope with heat loss by evaporative cooling and behavioral adaptations. An example of behavioral adaptation is that of a lizard lying in the sun on a hot rock in order to heat through conduction.

Endothermy

An endotherm is an animal that regulates its own body temperature, typically by keeping it at a constant level. To regulate body temperature, an organism may need to prevent heat gains in arid environments. Evaporation of water, either across respiratory surfaces or across the skin in those animals possessing sweat glands, helps in cooling body temperature to within the organism's tolerance range. Animals with a body covered by fur have limited ability to sweat, relying heavily on **panting** to increase evaporation of water across the moist surfaces of the lungs and the tongue and mouth. Mammals like cats, dogs and pigs, rely on panting or other means for thermal regulation and have sweat glands only in foot pads and snout. The sweat produced on pads of paws and on palms and soles mostly serves to increase friction and enhance grip. Birds also avoid overheating by gular fluttering, flapping the wings near the gular (throat) skin, similar to panting in mammals, since their thin skin has no sweat glands. Down feathers trap warm air acting as excellent insulators just as hair in mammals acts as a good insulator. Mammalian skin is much thicker than that of birds and often has a continuous layer of insulating fat beneath the dermis. In marine mammals, such as whales, or animals that live in very cold regions, such as the polar bears, this is called blubber. Dense coats found in desert endotherms also aid in preventing heat gain such as in the case of the camels.

A cold weather strategy is to temporarily decrease metabolic rate, decreasing the temperature difference between the animal and the air and thereby minimizing heat loss. Furthermore, having a lower metabolic rate is less energetically expensive. Many animals survive cold frosty nights through torpor, a short-term temporary drop in body temperature. Organisms when presented with the problem of regulating body temperature have not only behavioural, physiological, and structural adaptations but also a feedback system to trigger these adaptations to regulate temperature accordingly. The main features of this system are *stimulus, receptor, modulator, effector* and then the feedback of the newly adjusted temperature to the *stimulus*. This cyclical process aids in homeostasis.

Homeothermy Compared with Poikilothermy

Homeothermy and poikilothermy refer to how stable an organism's deep-body temperature is. Most endothermic organisms are homeothermic, like mammals. However, animals with facultative endothermy are often poikilothermic, meaning their temperature can vary considerably. Most fish are ectotherms, as most of their heat comes from the surrounding water. However, almost all fish are poikilothermic.

Vertebrates

By numerous observations upon humans and other animals, John Hunter showed that the essential difference between the so-called warm-blooded and cold-blooded animals lies in observed constancy of the temperature of the former, and the observed variability of the temperature of the latter. Almost all birds and mammals have a high temperature almost constant and independent of that of the surrounding air (homeothermy). Almost all other animals display a variation of body temperature, dependent on their surroundings (poikilothermy).

Brain Control

Thermoregulation in both ectotherms and endotherms is controlled mainly by the preoptic area of the anterior hypothalamus. Such homeostatic control is separate from the sensation of temperature.

In birds and Mammals

Kangaroo licking its arms to cool down on a very hot day

In cold environments, birds and mammals employ the following adaptations and strategies to minimize heat loss:

1. Using small smooth muscles (arrector pili in mammals), which are attached to feather or hair shafts; this distorts the surface of the skin making feather/hair shaft stand erect (called goose bumps or pimples) which slows the movement of air across the skin and minimizes heat loss.
2. Increasing body size to more easily maintain core body temperature (warm-blooded animals in cold climates tend to be larger than similar species in warmer climates.
3. Having the ability to store energy as fat for metabolism
4. Have shortened extremities
5. Have countercurrent blood flow in extremities - this is where the warm arterial blood travelling to the limb passes the cooler venous blood from the limb and heat is exchanged warming the venous blood and cooling the arterial (e.g., Arctic wolf or penguins)

In warm environments, birds and mammals employ the following adaptations and strategies to maximize heat loss:

1. Behavioural adaptations like living in burrows during the day and being nocturnal
2. Evaporative cooling by perspiration and panting
3. Storing fat reserves in one place (e.g., camel's hump) to avoid its insulating effect
4. Elongated, often vascularized extremities to conduct body heat to the air

In Humans

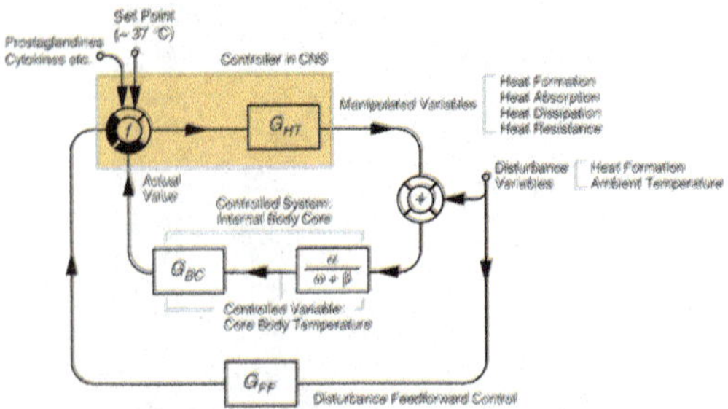

Simplified control circuit of human thermoregulation.

As in other mammals, thermoregulation is an important aspect of human homeostasis. Most body heat is generated in the deep organs, especially the liver, brain, and heart, and in contraction of skeletal muscles. Humans have been able to adapt to a great diversity of climates, including hot humid and hot arid. High temperatures pose serious stresses for the human body, placing it in great danger of injury or even death. For humans, adaptation to varying climatic conditions includes both physiological mechanisms resulting from evolution and behavioural mechanisms resulting from conscious cultural adaptations.

There are four avenues of heat loss: convection, conduction, radiation, and evaporation. If skin temperature is greater than that of the surroundings, the body can lose heat by radiation and conduction. But, if the temperature of the surroundings is greater than that of the skin, the body actually *gains* heat by radiation and conduction. In such conditions, the only means by which the body can rid itself of heat is by evaporation. So, when the surrounding temperature is higher than the skin temperature, anything that prevents adequate evaporation will cause the internal body temperature to rise. During intense physical activity (e.g. sports), evaporation becomes the main avenue of heat loss. Humidity affects thermoregulation by limiting sweat evaporation and thus heat loss.

A dog panting after exercise

In Plants

Thermogenesis occurs in the flowers of many plants in the Araceae family as well as in cycad cones. In addition, the sacred lotus (*Nelumbo nucifera*) is able to thermoregulate itself, remaining on average 20 °C (36 °F) above air temperature while flowering. Heat is produced by breaking down the starch that was stored in their roots, which requires the consumption of oxygen at a rate approaching that of a flying hummingbird.

One possible explanation for plant thermoregulation is to provide protection against cold temperature. For example, the skunk cabbage is not frost-resistant, yet it begins to grow and flower when there is still snow on the ground. Another theory is that thermogenicity helps attract pollinators, which is borne out by observations that heat production is accompanied by the arrival of beetles or flies.

Behavioral Temperature Regulation

Animals other than humans regulate and maintain their body temperature with physiological adjustments and behavior. Desert lizards are ectotherms and so unable to metabolically control their temperature but can do this by altering their location. They may do this, in the morning only by raising their head from its burrow and then exposing their entire body. By basking in the sun, the lizard absorbs solar heat. It may also absorb heat by conduction from heated rocks that have stored radiant solar energy. To lower their temperature, lizards exhibit varied behaviors. Sand seas, or ergs, produce up to 136 F (57.7C), and the sand lizard will hold its feet up in the air to cool down, seek cooler objects with which to contact, find shade or return to their burrow. They also go to their burrows to avoid cooling when the sun goes down or the temperature falls. Aquatic animals can also regulate their temperature behaviorally by changing their position in the thermal gradient.

During cold weather many animals increase their thermal inertia by huddling.

Animals also engage in kleptothermy in which they share or even steal each other's body warmth. In endotherms such as bats and birds (such as the mousebird and emperor penguin) it allows the sharing of body heat (particularly amongst juveniles). This allows the individuals to increase their thermal inertia (as with gigantothermy) and so reduce heat loss. Some ectotherms share burrows of ectotherms. Other animals exploit termite mounds.

Some animals living in cold environments maintain their body temperature by preventing heat loss. Their fur grows more densely to increase the amount of insulation. Some animals are regionally heterothermic and are able to allow their less insulated extremities to cool to temperatures much lower than their core temperature—nearly to 0 °C. This minimizes heat loss through less insulated body parts, like the legs, feet (or hooves), and nose.

An ostrich can keep its body temperature relatively constant, even though the environment can be very hot during the day and cold at night.

Hibernation, Estivation and Daily Torpor

To cope with limited food resources and low temperatures, some mammals hibernate during cold periods. To remain in "stasis" for long periods, these animals build up brown fat reserves and slow all body functions. True hibernators (e.g., groundhogs) keep their body temperatures low throughout hibernation whereas the core temperature of false hibernators (e.g., bears) varies; occasionally the animal may emerge from its den for brief periods. Some bats are true hibernators and rely upon a rapid, non-shivering thermogenesis of their brown fat deposit to bring them out of hibernation.

Estivation is similar to hibernation, however, it usually occurs in hot periods to allow animals to avoid high temperatures and desiccation. Both terrestrial and aquatic invertebrate and vertebrates enter into estivation. Examples include lady beetles (*Coccinellidae*), North American desert tortoises, crocodiles, salamanders, cane toads, and the water-holding frog

Daily torpor occurs in small endotherms like bats and hummingbirds, which temporarily reduces their high metabolic rates to conserve energy.

Variation in Animals

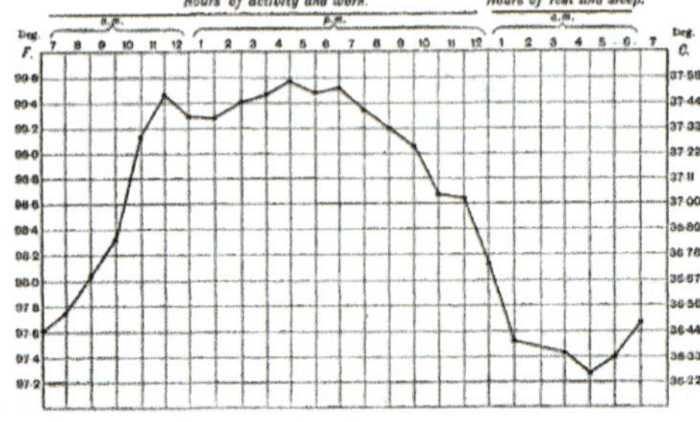

Chart showing diurnal variation in body temperature.

Normal Human Temperature

Previously, average oral temperature for healthy adults had been considered 37.0 °C (98.6 °F), while normal ranges are 36.1 °C (97.0 °F) to 37.8 °C (100.0 °F). In Poland and Russia, the temperature had been measured axillary. 36.6 °C was considered "ideal" temperature in these countries, while normal ranges are 36 °C to 36.9 °C.

Recent studies suggest that the average temperature for healthy adults is 36.8 °C (98.2 °F) (same result in three different studies). Variations (one standard deviation) from three other studies are:

- 36.4 - 37.1 °C (97.5 - 98.8 °F)
- 36.3 - 37.1 °C (97.3 - 98.8 °F) for males, 36.5 - 37.3 °C (97.7 - 99.1 °F) for females
- 36.6 - 37.3 °C (97.9 - 99.1 °F)

Measured temperature varies according to thermometer placement, with rectal temperature being 0.3-0.6 °C (0.5-1 °F) higher than oral temperature, while axillary temperature is 0.3-0.6 °C (0.5-1 °F) lower than oral temperature. The average difference between oral and axillary temperatures of Indian children aged 6–12 was found to be only 0.1 °C (standard deviation 0.2 °C), and the mean difference in Maltese children aged 4–14 between oral and axillary temperature was 0.56 °C, while the mean difference between rectal and axillary temperature for children under 4 years old was 0.38 °C.

Variations Due to Circadian Rhythms

In humans, a diurnal variation has been observed dependent on the periods of rest and activity, lowest at 11 p.m. to 3 a.m. and peaking at 10 a.m. to 6 p.m. Monkeys also have a well-marked and regular diurnal variation of body temperature that follows periods of rest and activity, and is not dependent on the incidence of day and night; nocturnal monkeys reach their highest body temperature at night and lowest during the day. Sutherland Simpson and J.J. Galbraith observed that all nocturnal animals and birds - whose periods of rest and activity are naturally reversed through habit and not from outside interference - experience their highest temperature during the natural period of activity (night) and lowest during the period of rest (day). Those diurnal temperatures can be reversed by reversing their daily routine.

In essence, the temperature curve of diurnal birds is similar to that of man and other homoeothermal animals, except that the maximum occurs earlier in the afternoon and the minimum earlier in the morning. Also, the curves obtained from rabbits, guinea pigs, and dogs were quite similar to those from man. These observations indicate that body temperature is partially regulated by circadian rhythms.

Variations Due to Women's Menstrual Cycles

During the follicular phase (which lasts from the first day of menstruation until the day of ovulation), the average basal body temperature in women ranges from 36.45 to 36.7 °C (97.6 to 98.1 °F). Within 24 hours of ovulation, women experience an elevation of 0.15 - 0.45 °C (0.2 - 0.9 °F) due to the increased metabolic rate caused by sharply elevated levels of progesterone. The basal body temperature ranges between 36.7 - 37.3 °C (98.1 - 99.2 °F) throughout the luteal phase, and drops down to pre-ovulatory levels within a few days of menstruation. Women can chart this phenomenon to determine whether and when they are ovulating, so as to aid conception or contraception.

Variations Due to Fever

Fever is a regulated elevation of the set point of core temperature in the hypothalamus, caused by circulating pyrogens produced by the immune system. To the subject, a rise in core temperature due to fever may result in feeling cold in an environment where people without fever do not.

Variations Due to Biofeedback

Some monks are known to practice Tummo, biofeedback meditation techniques, that allow them to raise their body temperatures substantially.

Low body Temperature Increases Lifespan

It has been theorised that low body temperature may increase lifespan. In 2006, it was reported that transgenic mice with a body temperature 0.3-0.5 C lower than normal mice lived longer than normal mice. This mechanism is due to overexpressing the uncoupling protein 2 in hypocretin neurons (Hcrt-UCP2), which elevated hypothalamic temperature, thus forcing the hypothalamus to lower body temperature. Lifespan was increased by 12% and 20% for males and females, respectively. The mice were fed *ad libitum*. The effects of such a genetic change in body temperature on longevity is more difficult to study in humans; in 2011, the UCP2 genetic alleles in humans were associated with obesity.

Limits Compatible with life

There are limits both of heat and cold that an endothermic animal can bear and other far wider limits that an ectothermic animal may endure and yet live. The effect of too extreme a cold is to decrease metabolism, and hence to lessen the production of heat. Both catabolic and anabolic pathways share in this metabolic depression, and, though less energy is used up, still less energy is generated. The effects of this diminished metabolism become telling on the central nervous system first, especially the brain and those parts concerning consciousness; both heart rate and respiration rate decrease;

judgment becomes impaired as drowsiness supervenes, becoming steadily deeper until the individual loses consciousness; without medical intervention, death by hypothermia quickly follows. Occasionally, however, convulsions may set in towards the end, and death is caused by asphyxia.

In experiments on cats performed by Sutherland Simpson and Percy T. Herring, the animals were unable to survive when rectal temperature fell below 16 °C. At this low temperature, respiration became increasingly feeble; heart-impulse usually continued after respiration had ceased, the beats becoming very irregular, appearing to cease, then beginning again. Death appeared to be mainly due to asphyxia, and the only certain sign that it had taken place was the loss of knee-jerks.

However, too high a temperature speeds up the metabolism of different tissues to such a rate that their metabolic capital is soon exhausted. Blood that is too warm produces dyspnea by exhausting the metabolic capital of the respiratory centre; heart rate is increased; the beats then become arrhythmic and eventually cease. The central nervous system is also profoundly affected by hyperthermia and delirium, and convulsions may set in. Consciousness may also be lost, propelling the person into a comatose condition. These changes can sometimes also be observed in patients suffering from an acute fever. Mammalian muscle becomes rigid with heat rigor at about 50 °C, with the sudden rigidity of the whole body rendering life impossible.

H.M. Vernon has done work on the death temperature and paralysis temperature (temperature of heat rigor) of various animals. He found that species of the same class showed very similar temperature values, those from the Amphibia examined being 38.5 °C, Fish 39 °C, Reptilia 45 °C, and various Molluscs 46 °C. Also, in the case of pelagic animals, he showed a relation between death temperature and the quantity of solid constituents of the body. In higher animals, however, his experiments tend to show that there is greater variation in both the chemical and physical characteristics of the protoplasm and, hence, greater variation in the extreme temperature compatible with life.

Arthropoda

The maximum temperatures tolerated by certain thermophilic arthropods exceeds the lethal temperatures for most vertebrates.

The most heat-resistant insects are three genera of desert ants recorded from three different parts of the world. The ants have developed a lifestyle of scavenging for short durations during the hottest hours of the day, in excess of 50 °C (122 °F) and often approaching 70 °C (158 °F), for the carcasses of insects and other forms of life which have succumbed to heat stress.

In April 2014, the South Californian mite *Paratarsotomus macropalpis* has been recorded as the world's fastest land animal relative to body length, at a speed of 322 body lengths per second. Besides the unusually great speed of the mites, the researchers were surprised to find the mites running at such speeds on concrete at temperatures up to 60 °C (140 °F), which is significant because this temperature is well above the lethal limit for the majority of animal species. In addition, the mites are able to stop and change direction very quickly.

References

- Tintinalli, Judith (2004). Emergency Medicine: A Comprehensive Study Guide, Sixth edition. McGraw-Hill Professional. p. 1181. ISBN 0-07-138875-3.
- Auerbach, [edited by] Paul S. (2007). Wilderness medicine. (5th ed.). St. Louis, Mo.: Elsevier Mosby. pp. Chapter 5. ISBN 978-0-323-03228-5.
- Marx, John (2010). Rosen's emergency medicine: concepts and clinical practice 7th edition. Philadelphia, PA: Mosby/Elsevier. p. 1868. ISBN 978-0-323-05472-0.
- James, William D.; Berger, Timothy G.; et al. (2006). Andrews' Diseases of the Skin: clinical Dermatology. Saunders Elsevier. ISBN 0-7216-2921-0.
- Robert Lacey and Danny Danziger August:The Year 1000: What Life Was Like at the Turn of the First Millennium Little, Brown, 2000 ISBN 0316511579
- Mistovich, Joseph; Haffen, Brent; Karren, Keith (2004). Prehospital Emergency Care. Upsaddle River, NJ: Pearson Education. p. 506. ISBN 0-13-049288-4.
- Swedan, Nadya Gabriele (2001). Women's Sports Medicine and Rehabilitation. Lippincott Williams & Wilkins. p. 149. ISBN 0-8342-1731-7.
- Sherwood, Van (1 May 1996). "Chapter 21: Most heat tolerant". Book of Insect Records. University of Florida. Retrieved 30 April 2014.
- Robertson, David (2012). Primer on the autonomic nervous system (3rd ed.). Amsterdam: Elsevier/AP. p. 288. ISBN 9780123865250.
- Bracker, Mark (2012). The 5-Minute Sports Medicine Consult (2 ed.). Lippincott Williams & Wilkins. p. 320. ISBN 9781451148121.

7

Applications of Cryobiology

Cryogenic temperatures have various uses. The application of cryobiological sciences for the preservation of cells and tissues is well known. It also plays an important part in the preservation and recovery of genetic codes of our long lost relatives. Some important applications of cryobiology are cloning, organ transplantation, IVF procedures and cryosurgery.

Cloning

Many organisms, including aspen trees, reproduce by cloning.

In biology, cloning is the process of producing similar populations of genetically identical individuals that occurs in nature when organisms such as bacteria, insects or plants reproduce asexually. Cloning in biotechnology refers to processes used to create copies of DNA fragments (molecular cloning), cells (cell cloning), or organisms. The term also refers to the production of multiple copies of a product such as digital media or software.

The term clone, invented by J. B. S. Haldane, is derived from the Ancient Greek word *klōn*, "twig", referring to the process whereby a new plant can be created from a twig. In horticulture, the spelling *clon* was used until the twentieth century; the final *e* came into use to indicate the vowel is a "long o" instead of a "short o". Since the term entered the popular lexicon in a more general context, the spelling *clone* has been used exclusively.

In botany, the term lusus was traditionally used.

Natural Cloning

Cloning is a natural form of reproduction that has allowed life forms to spread for more than 50 thousand years. It is the reproduction method used by plants, fungi, and bacteria, and is also the way that clonal colonies reproduce themselves. Examples of these organisms include blueberry plants, hazel trees, the Pando trees, the Kentucky coffeetree, *Myricas*, and the American sweetgum.

Molecular Cloning

Molecular cloning refers to the process of making multiple molecules. Cloning is commonly used to amplify DNA fragments containing whole genes, but it can also be used to amplify any DNA sequence such as promoters, non-coding sequences and randomly fragmented DNA. It is used in a wide array of biological experiments and practical applications ranging from genetic fingerprinting to large scale protein production. Occasionally, the term cloning is misleadingly used to refer to the identification of the chromosomal location of a gene associated with a particular phenotype of interest, such as in positional cloning. In practice, localization of the gene to a chromosome or genomic region does not necessarily enable one to isolate or amplify the relevant genomic sequence. To amplify any DNA sequence in a living organism, that sequence must be linked to an origin of replication, which is a sequence of DNA capable of directing the propagation of itself and any linked sequence. However, a number of other features are needed, and a variety of specialised cloning vectors (small piece of DNA into which a foreign DNA fragment can be inserted) exist that allow protein production, affinity tagging, single stranded RNA or DNA production and a host of other molecular biology tools.

Cloning of any DNA fragment essentially involves four steps

1. fragmentation - breaking apart a strand of DNA
2. ligation - gluing together pieces of DNA in a desired sequence
3. transfection - inserting the newly formed pieces of DNA into cells
4. screening/selection - selecting out the cells that were successfully transfected with the new DNA

Although these steps are invariable among cloning procedures a number of alternative routes can be selected; these are summarized as a *cloning strategy*.

Initially, the DNA of interest needs to be isolated to provide a DNA segment of suitable size. Subsequently, a ligation procedure is used where the amplified fragment is inserted into a vector (piece of DNA). The vector (which is frequently circular) is linearised using restriction enzymes, and incubated with the fragment of interest under appropriate conditions with an enzyme called DNA ligase. Following ligation the vector with the insert of interest is transfected into cells. A number of alternative techniques are available, such as chemical sensitivation of cells, electroporation, optical injection and biolistics. Finally, the transfected cells are cultured. As the aforementioned procedures are of particularly low efficiency, there is a need to identify the cells that have been successfully transfected with the vector construct containing the desired insertion sequence in the required orientation. Modern cloning vectors include selectable antibiotic resistance markers, which allow only cells in which the vector has been transfected, to grow. Additionally, the cloning vectors may contain colour selection markers, which provide blue/white screening (alpha-factor complementation) on X-gal medium. Nevertheless, these selection steps do not absolutely guarantee that the DNA insert is present in the cells obtained. Further investigation of the resulting colonies must be required to confirm that cloning was successful. This may be accomplished by means of PCR, restriction fragment analysis and/or DNA sequencing.

Cell Cloning

Cloning Unicellular Organisms

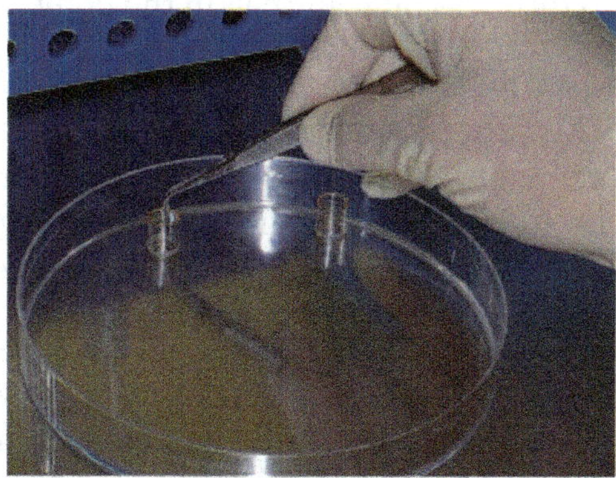
Cloning cell-line colonies using cloning rings

Cloning a cell means to derive a population of cells from a single cell. In the case of unicellular organisms such as bacteria and yeast, this process is remarkably simple and essentially only requires the inoculation of the appropriate medium. However, in the case of cell cultures from multi-cellular organisms, cell cloning is an arduous task as these cells will not readily grow in standard media.

A useful tissue culture technique used to clone distinct lineages of cell lines involves the use of cloning rings (cylinders). In this technique a single-cell suspension of cells that have been exposed to a mutagenic agent or drug used to drive selection is plated at high dilution to create isolated colonies, each arising from a single and potentially clonal distinct cell. At an early growth stage when colonies consist of only a few cells, sterile polystyrene rings (cloning rings), which have been dipped in grease, are placed over an individual colony and a small amount of trypsin is added. Cloned cells are collected from inside the ring and transferred to a new vessel for further growth.

Cloning Stem Cells

Somatic-cell nuclear transfer, known as SCNT, can also be used to create embryos for research or therapeutic purposes. The most likely purpose for this is to produce embryos for use in stem cell research. This process is also called "research cloning" or "therapeutic cloning." The goal is not to create cloned human beings (called "reproductive cloning"), but rather to harvest stem cells that can be used to study human development and to potentially treat disease. While a clonal human blastocyst has been created, stem cell lines are yet to be isolated from a clonal source.

Therapeutic cloning is achieved by creating embryonic stem cells in the hopes of treating diseases such as diabetes and Alzheimer's. The process begins by removing the nucleus (containing the DNA) from an egg cell and inserting a nucleus from the adult cell to be cloned. In the case of someone with Alzheimer's disease, the nucleus from a skin cell of that patient is placed into an empty egg. The reprogrammed cell begins to develop into an embryo because the egg reacts with the transferred nucleus. The embryo will become genetically identical to the patient. The embryo will then form a blastocyst which has the potential to form/become any cell in the body.

The reason why SCNT is used for cloning is because somatic cells can be easily acquired and cultured in the lab. This process can either add or delete specific genomes of farm animals. A key point to remember is that cloning is achieved when the oocyte maintains its normal functions and instead of using sperm and egg genomes to replicate, the oocyte is inserted into the donor's somatic cell nucleus. The oocyte will react on the somatic cell nucleus, the same way it would on sperm cells.

The process of cloning a particular farm animal using SCNT is relatively the same for all animals. The first step is to collect the somatic cells from the animal that will be cloned. The somatic cells could be used immediately or stored in the laboratory for later use. The hardest part of SCNT is removing maternal DNA from an oocyte at metaphase II. Once this has been done, the somatic nucleus can be inserted into an egg cytoplasm. This creates a one-cell embryo. The grouped somatic cell and egg cytoplasm are then introduced to an electrical current. This energy will hopefully allow the cloned embryo to begin development. The successfully developed embryos are then placed in surrogate recipients, such as a cow or sheep in the case of farm animals.

SCNT is seen as a good method for producing agriculture animals for food consumption. It successfully cloned sheep, cattle, goats, and pigs. Another benefit is SCNT is seen as a solution to clone endangered species that are on the verge of going extinct. However, stresses placed on both the egg cell and the introduced nucleus can be enormous, which led to a high loss in resulting cells in early research. For example, the cloned sheep Dolly was born after 277 eggs were used for SCNT, which created 29 viable embryos. Only three of these embryos survived until birth, and only one survived to adulthood. As the procedure could not be automated, and had to be performed manually under a microscope, SCNT was very resource intensive. The biochemistry involved in reprogramming the differentiated somatic cell nucleus and activating the recipient egg was also far from being well-understood. However, by

2014 researchers were reporting cloning success rates of seven to eight out of ten and in 2016, a Korean Company Sooam Biotech was reported to be producing 500 cloned embryos per day.

In SCNT, not all of the donor cell's genetic information is transferred, as the donor cell's mitochondria that contain their own mitochondrial DNA are left behind. The resulting hybrid cells retain those mitochondrial structures which originally belonged to the egg. As a consequence, clones such as Dolly that are born from SCNT are not perfect copies of the donor of the nucleus.

Organism Cloning

Organism cloning (also called reproductive cloning) refers to the procedure of creating a new multicellular organism, genetically identical to another. In essence this form of cloning is an asexual method of reproduction, where fertilization or inter-gamete contact does not take place. Asexual reproduction is a naturally occurring phenomenon in many species, including most plants and some insects. Scientists have made some major achievements with cloning, including the asexual reproduction of sheep and cows. There is a lot of ethical debate over whether or not cloning should be used. However, cloning, or asexual propagation, has been common practice in the horticultural world for hundreds of years.

Horticultural

The term *clone* is used in horticulture to refer to descendants of a single plant which were produced by vegetative reproduction or apomixis. Many horticultural plant cultivars are clones, having been derived from a single individual, multiplied by some process other than sexual reproduction. As an example, some European cultivars of grapes represent clones that have been propagated for over two millennia. Other examples are potato and banana. Grafting can be regarded as cloning, since all the shoots and branches coming from the graft are genetically a clone of a single individual, but this particular kind of cloning has not come under ethical scrutiny and is generally treated as an entirely different kind of operation.

Many trees, shrubs, vines, ferns and other herbaceous perennials form clonal colonies naturally. Parts of an individual plant may become detached by fragmentation and grow on to become separate clonal individuals. A common example is in the vegetative reproduction of moss and liverwort gametophyte clones by means of gemmae. Some vascular plants e.g. dandelion and certain viviparous grasses also form seeds asexually, termed apomixis, resulting in clonal populations of genetically identical individuals.

Parthenogenesis

Clonal derivation exists in nature in some animal species and is referred to as parthenogenesis (reproduction of an organism by itself without a mate). This is an asexual form of reproduction that is only found in females of some insects, crustaceans, nematodes, fish (for example the hammerhead shark), the Komodo dragon and lizards. The growth and development occurs without fertilization by a male. In plants, parthenogenesis means the development of an embryo from an unfertilized egg cell, and is a component process of apomixis. In species that use the XY sex-determination system, the offspring will always be female. An example is the little fire ant (*Wasmannia auropunctata*), which is native to Central and South America but has spread throughout many tropical environments.

Artificial Cloning of Organisms

Artificial cloning of organisms may also be called *reproductive cloning*.

First Moves

Hans Spemann, a German embryologist was awarded a Nobel Prize in Physiology or Medicine in 1935 for his discovery of the effect now known as embryonic induction, exercised by various parts of the embryo, that directs the development of groups of cells into particular tissues and organs. In 1928 he and his student, Hilde Mangold, were the first to perform somatic-cell nuclear transfer using amphibian embryos – one of the first moves towards cloning.

Methods

Reproductive cloning generally uses "somatic cell nuclear transfer" (SCNT) to create animals that are genetically identical. This process entails the transfer of a nucleus from a donor adult cell (somatic cell) to an egg from which the nucleus has been removed, or to a cell from a blastocyst from which the nucleus has been removed. If the egg begins to divide normally it is transferred into the uterus of the surrogate mother. Such clones are not strictly identical since the somatic cells may contain mutations in their nuclear DNA. Additionally, the mitochondria in the cytoplasm also contains DNA and during SCNT this mitochondrial DNA is wholly from the cytoplasmic donor's egg, thus the mitochondrial genome is not the same as that of the nucleus donor cell from which it was produced. This may have important implications for cross-species nuclear transfer in which nuclear-mitochondrial incompatibilities may lead to death.

Artificial *embryo splitting* or *embryo twinning*, a technique that creates monozygotic twins from a single embryo, is not considered in the same fashion as other methods of cloning. During that procedure, an donor embryo is split in two distinct embryos, that can then be transferred via embryo transfer. It is optimally performed at the 6- to 8-cell stage, where it can be used as an expansion of IVF to increase the number of available embryos. If both embryos are successful, it gives rise to monozygotic (identical) twins.

Dolly the Sheep

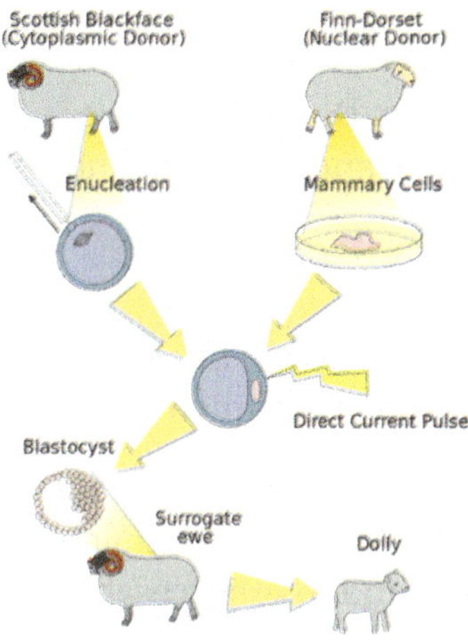

Dolly clone

Dolly, a Finn-Dorset ewe, was the first mammal to have been successfully cloned from an adult somatic cell. Dolly was formed by taking a cell from the udder of her 6-year old biological mother. Dolly's embryo was created by taking the cell and inserting it into a sheep ovum. It took 434 attempts before an embryo was successful. The embryo was then placed inside a female sheep that went through a normal pregnancy. She was cloned at the Roslin Institute in Scotland by British scientists Sir Ian Wilmut and Keith Campbell and lived there from her birth in 1996 until her death in 2003 when she was six. She was born on 5 July 1996 but not announced to the world until 22 February 1997. Her stuffed remains were placed at Edinburgh's Royal Museum, part of the National Museums of Scotland.

Dolly was publicly significant because the effort showed that genetic material from a specific adult cell, programmed to express only a distinct subset of its genes, can be reprogrammed to grow an entirely new organism. Before this demonstration, it had been shown by John Gurdon that nuclei from differentiated cells could give rise to an entire organism after transplantation into an enucleated egg. However, this concept was not yet demonstrated in a mammalian system.

The first mammalian cloning (resulting in Dolly the sheep) had a success rate of 29 embryos per 277 fertilized eggs, which produced three lambs at birth, one of which lived. In a bovine experiment involving 70 cloned calves, one-third of the calves died young. The first successfully cloned horse, Prometea, took 814 attempts. Notably, although the first clones were frogs, no adult cloned frog has yet been produced from a somatic adult nucleus donor cell.

There were early claims that Dolly the sheep had pathologies resembling accelerated aging. Scientists speculated that Dolly's death in 2003 was related to the shortening of telomeres, DNA-protein complexes that protect the end of linear chromosomes. However, other researchers, including Ian Wilmut who led the team that successfully cloned

Dolly, argue that Dolly's early death due to respiratory infection was unrelated to deficiencies with the cloning process. This idea that the nuclei have not irreversibly aged was shown in 2013 to be true for mice.

Dolly was named after performer Dolly Parton because the cells cloned to make her were from a mammary gland cell, and Parton is known for her ample cleavage.

Species Cloned

The modern cloning techniques involving nuclear transfer have been successfully performed on several species. Notable experiments include:

- Tadpole: (1952) Robert Briggs and Thomas J. King had successfully cloned northern leopard frogs: thirty-five complete embryos and twenty-seven tadpoles from one-hundred and four successful nuclear transfers.
- Carp: (1963) In China, embryologist Tong Dizhou produced the world's first cloned fish by inserting the DNA from a cell of a male carp into an egg from a female carp. He published the findings in a Chinese science journal.
- Mice: (1986) A mouse was successfully cloned from an early embryonic cell. Soviet scientists Chaylakhyan, Veprencev, Sviridova, and Nikitin had the mouse "Masha" cloned. Research was published in the magazine "Biofizika" volume XXXII, issue 5 of 1987.
- Sheep: Marked the first mammal being cloned (1984) from early embryonic cells by Steen Willadsen. Megan and Morag cloned from differentiated embryonic cells in June 1995 and Dolly the sheep from a somatic cell in 1996.
- Rhesus monkey: Tetra (January 2000) from embryo splitting
- Pig: the first cloned pigs (March 2000). By 2014, BGI in China was producing 500 cloned pigs a year to test new medicines.
- Gaur: (2001) was the first endangered species cloned.
- Cattle: Alpha and Beta (males, 2001) and (2005) Brazil
- Cat: CopyCat "CC" (female, late 2001), Little Nicky, 2004, was the first cat cloned for commercial reasons
- Rat: Ralph, the first cloned rat (2003)
- Mule: Idaho Gem, a john mule born 4 May 2003, was the first horse-family clone.
- Horse: Prometea, a Haflinger female born 28 May 2003, was the first horse clone.
- Dog: Snuppy, a male Afghan hound was the first cloned dog (2005).

- Wolf: Snuwolf and Snuwolffy, the first two cloned female wolves (2005).
- Water buffalo: Samrupa was the first cloned water buffalo. It was born on 6 February 2009, at India's Karnal National Diary Research Institute but died five days later due to lung infection.
- Pyrenean ibex (2009) was the first extinct animal to be cloned back to life; the clone lived for seven minutes before dying of lung defects.
- Camel: (2009) Injaz, is the first cloned camel.
- Pashmina goat: (2012) Noori, is the first cloned pashmina goat. Scientists at the faculty of veterinary sciences and animal husbandry of Sher-e-Kashmir University of Agricultural Sciences and Technology of Kashmir successfully cloned the first Pashmina goat (Noori) using the advanced reproductive techniques under the leadership of Riaz Ahmad Shah.
- Gastric brooding frog: (2013) The gastric brooding frog, *Rheobatrachus silus*, thought to have been extinct since 1983 was cloned in Australia, although the embryos died after a few days.

Human Cloning

Human cloning is the creation of a genetically identical copy of a human. The term is generally used to refer to artificial human cloning, which is the reproduction of human cells and tissues. It does not refer to the natural conception and delivery of identical twins. The possibility of human cloning has raised controversies. These ethical concerns have prompted several nations to pass legislature regarding human cloning and its legality.

Two commonly discussed types of theoretical human cloning are *therapeutic cloning* and *reproductive cloning*. Therapeutic cloning would involve cloning cells from a human for use in medicine and transplants, and is an active area of research, but is not in medical practice anywhere in the world, as of 2014. Two common methods of therapeutic cloning that are being researched are somatic-cell nuclear transfer and, more recently, pluripotent stem cell induction. Reproductive cloning would involve making an entire cloned human, instead of just specific cells or tissues.

Ethical Issues of Cloning

There are a variety of ethical positions regarding the possibilities of cloning, especially human cloning. While many of these views are religious in origin, the questions raised by cloning are faced by secular perspectives as well. Perspectives on human cloning are theoretical, as human therapeutic and reproductive cloning are not commercially used; animals are currently cloned in laboratories and in livestock production.

Advocates support development of therapeutic cloning in order to generate tissues and whole organs to treat patients who otherwise cannot obtain transplants, to avoid the need for immunosuppressive drugs, and to stave off the effects of aging. Advocates for reproductive cloning believe that parents who cannot otherwise procreate should have access to the technology.

Opponents of cloning have concerns that technology is not yet developed enough to be safe and that it could be prone to abuse (leading to the generation of humans from whom organs and tissues would be harvested), as well as concerns about how cloned individuals could integrate with families and with society at large.

Religious groups are divided, with some opposing the technology as usurping "God's place" and, to the extent embryos are used, destroying a human life; others support therapeutic cloning's potential life-saving benefits.

Cloning of animals is opposed by animal-groups due to the number of cloned animals that suffer from malformations before they die, and while food from cloned animals has been approved by the US FDA, its use is opposed by groups concerned about food safety.

Cloning Extinct and Endangered Species

Cloning, or more precisely, the reconstruction of functional DNA from extinct species has, for decades, been a dream. Possible implications of this were dramatized in the 1984 novel *Carnosaur* and the 1990 novel *Jurassic Park*. The best current cloning techniques have an average success rate of 9.4 percent (and as high as 25 percent) when working with familiar species such as mice, while cloning wild animals is usually less than 1 percent successful. Several tissue banks have come into existence, including the "Frozen Zoo" at the San Diego Zoo, to store frozen tissue from the world's rarest and most endangered species.

In 2001, a cow named Bessie gave birth to a cloned Asian gaur, an endangered species, but the calf died after two days. In 2003, a banteng was successfully cloned, followed by three African wildcats from a thawed frozen embryo. These successes provided hope that similar techniques (using surrogate mothers of another species) might be used to clone extinct species. Anticipating this possibility, tissue samples from the last *bucardo* (Pyrenean ibex) were frozen in liquid nitrogen immediately after it died in 2000. Researchers are also considering cloning endangered species such as the giant panda and cheetah.

In 2002, geneticists at the Australian Museum announced that they had replicated DNA of the thylacine (Tasmanian tiger), at the time extinct for about 65 years, using polymerase chain reaction. However, on 15 February 2005 the museum announced that it was stopping the project after tests showed the specimens' DNA had been too badly degraded by the (ethanol) preservative. On 15 May 2005 it was announced that the thylacine project would be revived, with new participation from researchers in New South Wales and Victoria.

In January 2009, for the first time, an extinct animal, the Pyrenean ibex mentioned above was cloned, at the Centre of Food Technology and Research of Aragon, using the preserved frozen cell nucleus of the skin samples from 2001 and domestic goat egg-cells. The ibex died shortly after birth due to physical defects in its lungs.

One of the most anticipated targets for cloning was once the woolly mammoth, but attempts to extract DNA from frozen mammoths have been unsuccessful, though a joint Russo-Japanese team is currently working toward this goal. In January 2011, it was reported by Yomiuri Shimbun that a team of scientists headed by Akira Iritani of Kyoto University had built upon research by Dr. Wakayama, saying that they will extract DNA from a mammoth carcass that had been preserved in a Russian laboratory and insert it into the egg cells of an African elephant in hopes of producing a mammoth embryo. The researchers said they hoped to produce a baby mammoth within six years. It was noted, however that the result, if possible, would be an elephant-mammoth hybrid rather than a true mammoth. Another problem is the survival of the reconstructed mammoth: ruminants rely on a symbiosis with specific microbiota in their stomachs for digestion.

Scientists at the University of Newcastle and University of New South Wales announced in March 2013 that the very recently extinct gastric-brooding frog would be the subject of a cloning attempt to resurrect the species.

Many such “de-extinction” projects are described in the Long Now Foundation’s Revive and Restore Project.

Lifespan

After an eight-year project involving the use of a pioneering cloning technique, Japanese researchers created 25 generations of healthy cloned mice with normal lifespans, demonstrating that clones are not intrinsically shorter-lived than naturally born animals.

In a detailed study released in 2016 and less detailed studies by others suggest that once cloned animals get past the first month or two of life they are generally healthy. However, early pregnancy loss and neonatal losses are still greater with cloning than natural conception or assisted reproduction (IVF). Current research endeavors are attempting to overcome this problem.

In Popular Culture

In an article in the 8 November 1993 article of *Time*, cloning was portrayed in a negative way, modifying Michelangelo’s *Creation of Adam* to depict Adam with five identical hands. *Newsweek*’s 10 March 1997 issue also critiqued the ethics of human cloning, and included a graphic depicting identical babies in beakers.

Cloning is a recurring theme in a wide variety of contemporary science fiction, ranging from action films such as *Jurassic Park* (1993), *The 6th Day* (2000), *Resident Evil*

(2002), *Star Wars* (2002) and *The Island* (2005), to comedies such as Woody Allen's 1973 film *Sleeper*.

Science fiction has used cloning, most commonly and specifically human cloning, due to the fact that it brings up controversial questions of identity. *A Number* is a 2002 play by English playwright Caryl Churchill which addresses the subject of human cloning and identity, especially nature and nurture. The story, set in the near future, is structured around the conflict between a father (Salter) and his sons (Bernard 1, Bernard 2, and Michael Black) – two of whom are clones of the first one. *A Number* was adapted by Caryl Churchill for television, in a co-production between the BBC and HBO Films.

A recurring sub-theme of cloning fiction is the use of clones as a supply of organs for transplantation. The 2005 Kazuo Ishiguro novel *Never Let Me Go* and the 2010 film adaption are set in an alternate history in which cloned humans are created for the sole purpose of providing organ donations to naturally born humans, despite the fact that they are fully sentient and self-aware. The 2005 film *The Island* revolves around a similar plot, with the exception that the clones are unaware of the reason for their existence.

The use of human cloning for military purposes has also been explored in several works. *Star Wars* portrays human cloning in *Clone Wars*.

The exploitation of human clones for dangerous and undesirable work was examined in the 2009 British science fiction film *Moon*. In the futuristic novel *Cloud Atlas* and subsequent film, one of the story lines focuses on a genetically-engineered fabricant clone named Sonmi~451 who is one of millions raised in an artificial "wombtank," destined to serve from birth. She is one of thousands of clones created for manual and emotional labor; Sonmi herself works as a server in a restaurant. She later discovers that the sole source of food for clones, called 'Soap', is manufactured from the clones themselves.

Cloning has been used in fiction as a way of recreating historical figures. In the 1976 Ira Levin novel *The Boys from Brazil* and its 1978 film adaptation, Josef Mengele uses cloning to create copies of Adolf Hitler.

In 2012, a Japanese television show named "Bunshin" was created. The story's main character, Mariko, is a woman studying child welfare in Hokkaido. She grew up always doubtful about the love from her mother, who looked nothing like her and who died nine years before. One day, she finds some of her mother's belongings at a relative's house, and heads to Tokyo to seek out the truth behind her birth. She later discovered that she was a clone.

In the 2013 television show *Orphan Black*, cloning is used as a scientific study on the behavioral adaptation of the clones. In a similar vein, the book *The Double* by Nobel Prize winner José Saramago explores the emotional experience of a man who discovers that he is a clone.

Molecular Cloning

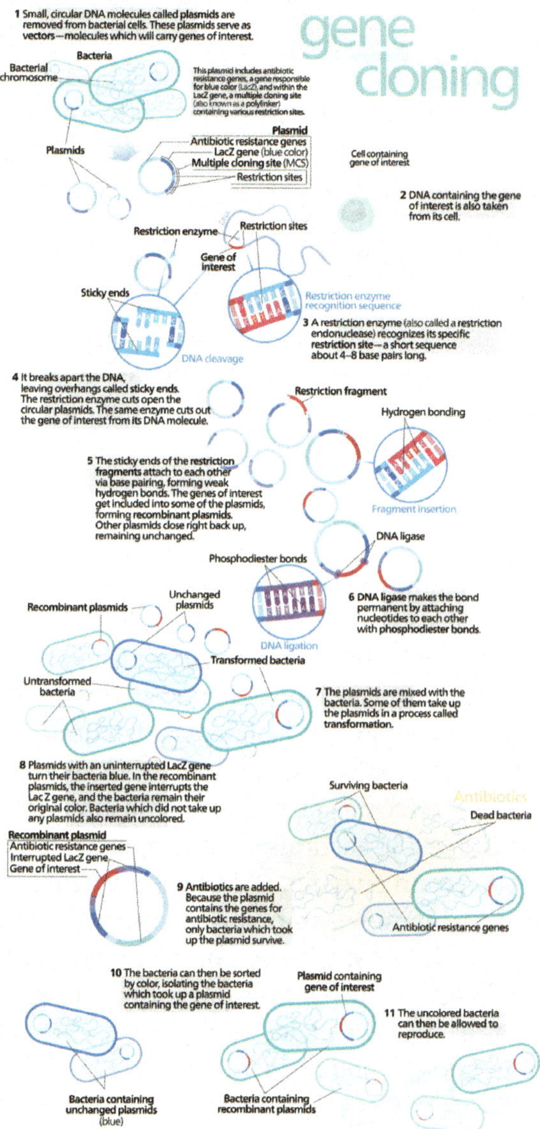

gene cloning
1 Small, circular DNA molecules called plasmids are removed from bacterial cells. These plasmids serve as vectors—molecules which will carry genes of interest.
Bacteria
Bacterial chromosome
This plasmid includes antibiotic resistance genes, a gene responsible for blue color (LacZ), and within the LacZ gene, a multiple cloning site (also known as a polylinker) containing various restriction sites.
Plasmid
Antibiotic resistance genes
LacZ gene (blue color)
Multiple cloning site (MCS)
Restriction sites
Plasmids
Cell containing gene of interest
2 DNA containing the gene of interest is also taken from its cell.
Restriction enzyme
Restriction sites
Gene of interest
Sticky ends
Restriction enzyme recognition sequence
3 A restriction enzyme (also called a restriction endonuclease) recognizes its specific restriction site—a short sequence about 4–8 base pairs long.
DNA cleavage
4 It breaks apart the DNA, leaving overhangs called sticky ends. The restriction enzyme cuts open the circular plasmids. The same enzyme cuts out the gene of interest from its DNA molecule.
Restriction fragment
Hydrogen bonding
5 The sticky ends of the restriction fragments attach to each other via base pairing, forming weak hydrogen bonds. The genes of interest get included into some of the plasmids, forming recombinant plasmids. Other plasmids close right back up, remaining unchanged.
Fragment insertion
DNA ligase
Phosphodiester bonds
6 DNA ligase makes the bond permanent by attaching nucleotides to each other with phosphodiester bonds.
DNA ligation
Recombinant plasmids
Unchanged plasmids
Transformed bacteria
Untransformed bacteria
7 The plasmids are mixed with the bacteria. Some of them take up the plasmids in a process called transformation.
8 Plasmids with an uninterrupted LacZ gene turn their bacteria blue. In the recombinant plasmids, the inserted gene interrupts the Lac Z gene, and the bacteria remain their original color. Bacteria which did not take up any plasmids also remain uncolored.
Surviving bacteria
Antibiotics
Dead bacteria
Recombinant plasmid
Antibiotic resistance genes
Interrupted LacZ gene
Gene of interest
9 Antibiotics are added. Because the plasmid contains the genes for antibiotic resistance, only bacteria which took up the plasmid survive.
Antibiotic resistance genes
10 The bacteria can then be sorted by color, isolating the bacteria which took up a plasmid containing the gene of interest.
Plasmid containing gene of interest
11 The uncolored bacteria can then be allowed to reproduce.
Bacteria containing unchanged plasmids (blue)
Bacteria containing recombinant plasmids

Diagram of molecular cloning using bacteria and plasmids.

Molecular cloning is a set of experimental methods in molecular biology that are used to assemble recombinant DNA molecules and to direct their replication within host organisms. The use of the word *cloning* refers to the fact that the method involves the replication of one molecule to produce a population of cells with identical DNA molecules. Molecular cloning generally uses DNA sequences from two different organisms: the species that is the source of the DNA to be cloned, and the species that will serve as the living host for replication of the recombinant DNA. Molecular cloning methods are central to many contemporary areas of modern biology and medicine.

In a conventional molecular cloning experiment, the DNA to be cloned is obtained from an organism of interest, then treated with enzymes in the test tube to generate smaller DNA fragments. Subsequently, these fragments are then combined with vector DNA to generate recombinant DNA molecules. The recombinant DNA is then introduced into a host organism (typically an easy-to-grow, benign, laboratory strain of *E. coli* bacteria). This will generate a population of organisms in which recombinant DNA molecules are replicated along with the host DNA. Because they contain foreign DNA fragments, these are transgenic or genetically modified microorganisms (GMO). This process takes advantage of the fact that a single bacterial cell can be induced to take up and replicate a single recombinant DNA molecule. This single cell can then be expanded exponentially to generate a large amount of bacteria, each of which contain copies of the original recombinant molecule. Thus, both the resulting bacterial population, and the recombinant DNA molecule, are commonly referred to as "clones". Strictly speaking, *recombinant DNA* refers to DNA molecules, while *molecular cloning* refers to the experimental methods used to assemble them.

History of Molecular Cloning

Prior to the 1970s, our understanding of genetics and molecular biology was severely hampered by an inability to isolate and study individual genes from complex organisms. This changed dramatically with the advent of molecular cloning methods. Microbiologists, seeking to understand the molecular mechanisms through which bacteria restricted the growth of bacteriophage, isolated restriction endonucleases, enzymes that could cleave DNA molecules only when specific DNA sequences were encountered. They showed that restriction enzymes cleaved chromosome-length DNA molecules at specific locations, and that specific sections of the larger molecule could be purified by size fractionation. Using a second enzyme, DNA ligase, fragments generated by restriction enzymes could be joined in new combinations, termed recombinant DNA. By recombining DNA segments of interest with vector DNA, such as bacteriophage or plasmids, which naturally replicate inside bacteria, large quantities of purified recombinant DNA molecules could be produced in bacterial cultures. The first recombinant DNA molecules were generated and studied in 1972.

Overview

Molecular cloning takes advantage of the fact that the chemical structure of DNA is fundamentally the same in all living organisms. Therefore, if any segment of DNA from any organism is inserted into a DNA segment containing the molecular sequences required for DNA replication, and the resulting recombinant DNA is introduced into the organism from which the replication sequences were obtained, then the foreign DNA will be replicated along with the host cell's DNA in the transgenic organism.

Molecular cloning is similar to polymerase chain reaction (PCR) in that it permits the replication of DNA sequence. The fundamental difference between the two methods is that molecular cloning involves replication of the DNA in a living microorganism, while PCR replicates DNA in an *in vitro* solution, free of living cells.

Steps in Molecular Cloning

In standard molecular cloning experiments, the cloning of any DNA fragment essentially involves seven steps: (1) Choice of host organism and cloning vector, (2) Preparation of vector DNA, (3) Preparation of DNA to be cloned, (4) Creation of recombinant DNA, (5) Introduction of recombinant DNA into host organism, (6) Selection of organisms containing recombinant DNA, (7) Screening for clones with desired DNA inserts and biological properties.

Although the detailed planning of the cloning can be done in any text editor, together with online utilities for e.g. PCR primer design, dedicated software exist for the purpose. Software for the purpose include for example ApE (open source), DNAStrider (open source), Serial Cloner (gratis) and Collagene (open source).

Choice of Host Organism and Cloning Vector

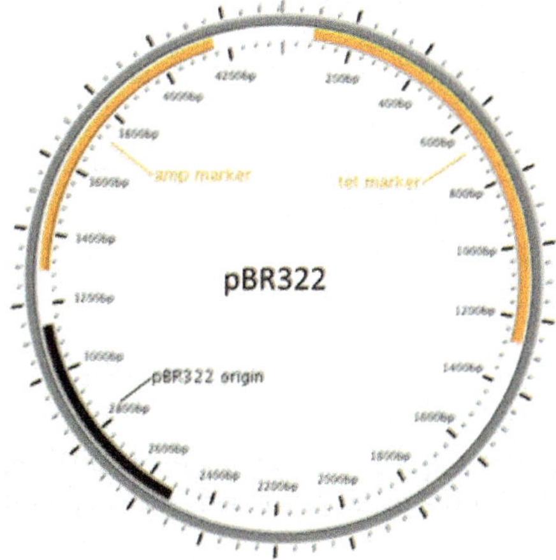

Diagram of a commonly used cloning plasmid; pBR322. It's a circular piece of DNA 4361 bases long. Two antibiotic resistance genes are present, conferring resistance to ampicillin and tetracycline, and an origin of replication that the host uses to replicate the DNA.

Although a very large number of host organisms and molecular cloning vectors are in use, the great majority of molecular cloning experiments begin with a laboratory strain of the bacterium *E. coli* (*Escherichia coli*) and a plasmid cloning vector. *E. coli* and plasmid vectors are in common use because they are technically sophisticated, versatile, widely available, and offer rapid growth of recombinant organisms with minimal equipment. If the DNA to be cloned is exceptionally large (hundreds of thousands to millions of base pairs), then a bacterial artificial chromosome or yeast artificial chromosome vector is often chosen.

Specialized applications may call for specialized host-vector systems. For example, if the experimentalists wish to harvest a particular protein from the recombinant organism, then an expression vector is chosen that contains appropriate signals for transcription and translation in the desired host organism. Alternatively, if replication of the DNA in different species is desired (for example, transfer of DNA from bacteria to plants), then a multiple host range vector (also termed shuttle vector) may be selected. In practice, however, specialized molecular cloning experiments usually begin with cloning into a bacterial plasmid, followed by subcloning into a specialized vector.

Whatever combination of host and vector are used, the vector almost always contains four DNA segments that are critically important to its function and experimental utility:

1. DNA *replication origin* is necessary for the vector (and its linked recombinant sequences) to replicate inside the host organism
2. one or more unique *restriction endonuclease recognition sites* to serves as sites where foreign DNA may be introduced
3. a *selectable genetic marker* gene that can be used to enable the survival of cells that have taken up vector sequences
4. a *tag* gene that can be used to screen for cells containing the foreign DNA

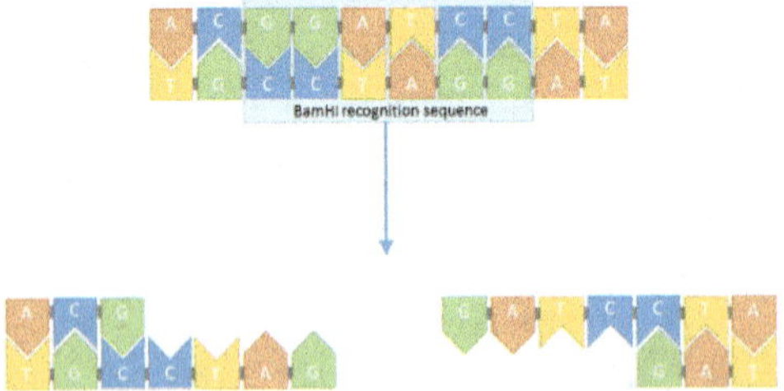

Cleavage of a DNA sequence containing the BamHI restriction site. The DNA is cleaved at the palindromic sequence to produce 'sticky ends'.

Preparation of Vector DNA

The cloning vector is treated with a restriction endonuclease to cleave the DNA at the site where foreign DNA will be inserted. The restriction enzyme is chosen to generate a configuration at the cleavage site that is compatible with the ends of the foreign DNA Typically, this is done by cleaving the vector DNA and foreign DNA with the same restriction enzyme, for example EcoRI. Most modern vectors contain a variety of convenient cleavage sites that are unique within the vector molecule (so that the vector can only be cleaved at a single site) and are located within a gene (frequently beta-galactosidase) whose inactivation can be used to distinguish recombinant from non-recombinant organisms at a later step in the process. To improve the ratio of recombinant to non-recombinant organisms, the cleaved vector may be treated with an enzyme (alkaline phosphatase) that dephosphorylates the vector ends. Vector molecules with dephosphorylated ends are unable to replicate, and replication can only be restored if foreign DNA is integrated into the cleavage site.

Preparation of DNA to be Cloned

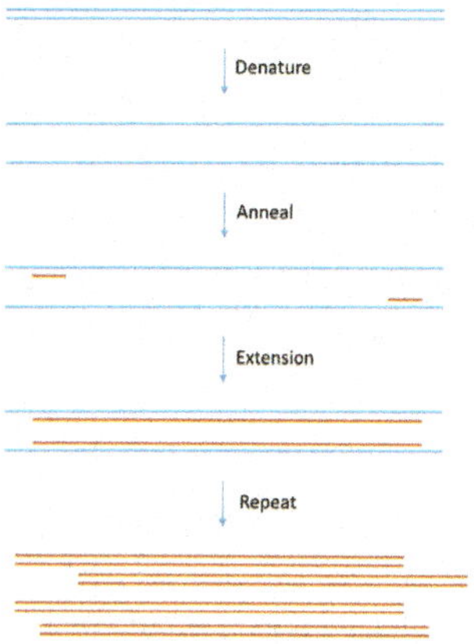

Denature

Anneal

Extension

Repeat

DNA for cloning is most commonly produced using PCR. Template DNA is mixed with bases (the building blocks of DNA), primers (short pieces of complementary single stranded DNA) and a DNA polymerase enzyme that builds the DNA chain. The mix goes through cycles of heating and cooling to produce large quantities of copied DNA.

For cloning of genomic DNA, the DNA to be cloned is extracted from the organism of interest. Virtually any tissue source can be used (even tissues from extinct animals), as long as the DNA is not extensively degraded. The DNA is then purified using simple methods to remove contaminating proteins (extraction with phenol), RNA (ribonuclease) and smaller molecules (precipitation and/or chromatography). Polymerase chain reaction (PCR) methods are often used for amplification of specific DNA or RNA (RT-PCR) sequences prior to molecular cloning.

DNA for cloning experiments may also be obtained from RNA using reverse transcriptase (complementary DNA or cDNA cloning), or in the form of synthetic DNA (artificial gene synthesis). cDNA cloning is usually used to obtain clones representative of the mRNA population of the cells of interest, while synthetic DNA is used to obtain any precise sequence defined by the designer.

The purified DNA is then treated with a restriction enzyme to generate fragments with ends capable of being linked to those of the vector. If necessary, short double-stranded segments of DNA (*linkers*) containing desired restriction sites may be added to create end structures that are compatible with the vector.

Creation of Recombinant DNA with DNA Ligase

The creation of recombinant DNA is in many ways the simplest step of the molecular cloning process. DNA prepared from the vector and foreign source are simply mixed together at appropriate concentrations and exposed to an enzyme (DNA ligase) that covalently links the ends together. This joining reaction is often termed ligation. The resulting DNA mixture containing randomly joined ends is then ready for introduction into the host organism.

DNA ligase only recognizes and acts on the ends of linear DNA molecules, usually resulting in a complex mixture of DNA molecules with randomly joined ends. The desired products (vector DNA covalently linked to foreign DNA) will be present, but other sequences (e.g. foreign DNA linked to itself, vector DNA linked to itself and higher-order combinations of vector and foreign DNA) are also usually present. This complex mixture is sorted out in subsequent steps of the cloning process, after the DNA mixture is introduced into cells.

Introduction of Recombinant DNA into Host Organism

The DNA mixture, previously manipulated in vitro, is moved back into a living cell, referred to as the host organism. The methods used to get DNA into cells are varied, and the name applied to this step in the molecular cloning process will often depend upon the experimental method that is chosen (e.g. transformation, transduction, transfection, electroporation).

When microorganisms are able to take up and replicate DNA from their local environment, the process is termed transformation, and cells that are in a physiological state such that they can take up DNA are said to be competent. In mammalian cell culture, the analogous process of introducing DNA into cells is commonly termed transfection. Both transformation and transfection usually require preparation of the cells through a special growth regime and chemical treatment process that will vary with the specific species and cell types that are used.

Electroporation uses high voltage electrical pulses to translocate DNA across the cell membrane (and cell wall, if present). In contrast, transduction involves the packaging of DNA into virus-derived particles, and using these virus-like particles to introduce the encapsulated DNA into the cell through a process resembling viral infection. Although electroporation and transduction are highly specialized methods, they may be the most efficient methods to move DNA into cells.

Selection of Organisms Containing Vector Sequences

Whichever method is used, the introduction of recombinant DNA into the chosen host organism is usually a low efficiency process; that is, only a small fraction of the cells will actually take up DNA. Experimental scientists deal with this issue through a step of artificial genetic selection, in which cells that have not taken up DNA are selectively killed, and only those cells that can actively replicate DNA containing the selectable marker gene encoded by the vector are able to survive.

When bacterial cells are used as host organisms, the selectable marker is usually a gene that confers resistance to an antibiotic that would otherwise kill the cells, typically ampicillin. Cells harboring the plasmid will survive when exposed to the antibiotic, while those that have failed to take up plasmid sequences will die. When mammalian cells (e.g. human or mouse cells) are used, a similar strategy is used, except that the marker gene (in this case typically encoded as part of the kanMX cassette) confers resistance to the antibiotic Geneticin.

Screening for Clones with Desired DNA Inserts and Biological Properties

Modern bacterial cloning vectors (e.g. pUC19 and later derivatives including the pGEM vectors) use the blue-white screening system to distinguish colonies (clones) of transgenic cells from those that contain the parental vector (i.e. vector DNA with no recombinant sequence inserted). In these vectors, foreign DNA is inserted into a sequence that encodes an essential part of beta-galactosidase, an enzyme whose activity results in formation of a blue-colored colony on the culture medium that is used for this work. Insertion of the foreign DNA into the beta-galactosidase coding sequence disables the function of the enzyme, so that colonies containing transformed DNA remain colorless (white). Therefore, experimentalists are easily able to identify and conduct further studies on transgenic bacterial clones, while ignoring those that do not contain recombinant DNA.

The total population of individual clones obtained in a molecular cloning experiment is often termed a DNA library. Libraries may be highly complex (as when cloning complete genomic DNA from an organism) or relatively simple (as when moving a previously cloned DNA fragment into a different plasmid), but it is almost always necessary to examine a number of different clones to be sure that the desired DNA construct is obtained. This may be accomplished through a very wide range of experimental methods, including the use of nucleic acid hybridizations, antibody probes, polymerase chain reaction, restriction fragment analysis and/or DNA sequencing.

Applications of Molecular Cloning

Molecular cloning provides scientists with an essentially unlimited quantity of any individual DNA segments derived from any genome. This material can be used for a wide range of purposes, including those in both basic and applied biological science. A few of the more important applications are summarized here.

Genome Organization and Gene Expression

Molecular cloning has led directly to the elucidation of the complete DNA sequence of the genomes of a very large number of species and to an exploration of genetic diversity within individual species, work that has been done mostly by determining the DNA sequence of large numbers of randomly cloned fragments of the genome, and assembling the overlapping sequences.

At the level of individual genes, molecular clones are used to generate probes that are used for examining how genes are expressed, and how that expression is related to other processes in biology, including the metabolic environment, extracellular signals, development, learning, senescence and cell death. Cloned genes can also provide tools to examine the biological function and importance of individual genes, by allowing investigators to inactivate the genes, or make more subtle mutations using regional mutagenesis or site-directed mutagenesis.

Production of Recombinant Proteins

Obtaining the molecular clone of a gene can lead to the development of organisms that produce the protein product of the cloned genes, termed a recombinant protein. In practice, it is frequently more difficult to develop an organism that produces an active form of the recombinant protein in desirable quantities than it is to clone the gene. This is because the molecular signals for gene expression are complex and variable, and because protein folding, stability and transport can be very challenging.

Many useful proteins are currently available as recombinant products. These include--(1) medically useful proteins whose administration can correct a defective or poorly expressed gene (e.g. recombinant factor VIII, a blood-clotting factor deficient in some forms of hemophilia, and recombinant insulin, used to treat some forms of diabetes), (2) proteins that can be administered to assist in a life-threatening emergency (e.g. tissue plasminogen activator, used to treat strokes), (3) recombinant subunit vaccines, in which a purified protein can be used to immunize patients against infectious diseases, without exposing them to the infectious agent itself (e.g. hepatitis B vaccine), and (4) recombinant proteins as standard material for diagnostic laboratory tests.

Transgenic Organisms

Once characterized and manipulated to provide signals for appropriate expression, cloned genes may be inserted into organisms, generating transgenic organisms, also termed genetically modified organisms (GMOs). Although most GMOs are generated for purposes of basic biological research , a number of GMOs have been developed for commercial use, ranging from animals and plants that produce pharmaceuticals or other compounds (pharming), herbicide-resistant crop plants, and fluorescent tropical fish (GloFish) for home entertainment.

Gene Therapy

Gene therapy involves supplying a functional gene to cells lacking that function, with the aim of correcting a genetic disorder or acquired disease. Gene therapy can be broadly divided into two categories. The first is alteration of germ cells, that is, sperm or eggs, which results in a permanent genetic change for the whole organism and subsequent generations. This "germ line gene therapy" is considered by many to be unethical in human beings. The second type of gene therapy, "somatic cell gene therapy", is analogous to an organ transplant. In this case, one or more specific tissues are targeted by direct treatment or by removal of the tissue, addition of the therapeutic gene or genes in the laboratory, and return of the treated cells to the patient. Clinical trials of somatic cell gene therapy began in the late 1990s, mostly for the treatment of cancers and blood, liver, and lung disorders.

Despite a great deal of publicity and promises, the history of human gene therapy has been characterized by relatively limited success. The effect of introducing a gene into cells often promotes only partial and/or transient relief from the symptoms of the disease being treated. Some gene therapy trial patients have suffered adverse consequences of the treatment itself, including deaths. In some cases, the adverse effects result from disruption of essential genes within the patient's genome by insertional inactivation. In others, viral vectors used for gene therapy have been contaminated with infectious virus. Nevertheless, gene therapy is still held to be a promising future area of medicine, and is an area where there is a significant level of research and development activity.

Organ Transplantation

Organ transplantation is the moving of an organ from one body to another or from a donor site to another location on the person's own body, to replace the recipient's damaged or absent organ. Organs and/or tissues that are transplanted within the same person's body are called autografts. Transplants that are recently performed between two subjects of the same species are called allografts. Allografts can either be from a living or cadaveric source.

Organs that can be transplanted are the heart, kidneys, liver, lungs, pancreas, intestine, and thymus. Tissues include bones, tendons (both referred to as musculoskeletal grafts), cornea, skin, heart valves, nerves and veins. Worldwide, the kidneys are the most commonly transplanted organs, followed by the liver and then the heart. Cornea and musculoskeletal grafts are the most commonly transplanted tissues; these outnumber organ transplants by more than tenfold.

Organ donors may be living, brain dead, or dead via circulatory death. Tissue may be recovered from donors who die of circulatory death, as well as of brain death – up to 24 hours past the cessation of heartbeat. Unlike organs, most tissues (with the exception of corneas) can be preserved and stored for up to five years, meaning they can be "banked". Transplantation raises a number of bioethical issues, including the definition of death, when and how consent should be given for an organ to be transplanted, and payment for organs for transplantation. Other ethical issues include transplantation tourism and more broadly the socio-economic context in which organ procurement or transplantation may occur. A particular problem is organ trafficking. Some organs, such as the brain, cannot be transplanted.

Transplantation medicine is one of the most challenging and complex areas of modern medicine. Some of the key areas for medical management are the problems of transplant rejection, during which the body has an immune response to the transplanted organ, possibly leading to transplant failure and the need to immediately remove the organ from the recipient. When possible, transplant rejection can be reduced through serotyping to determine the most appropriate donor-recipient match and through the use of immunosuppressant drugs.

Types of Transplant

Autograft

Autografts are the transplant of tissue to the same person. Sometimes this is done with surplus tissue, tissue that can regenerate, or tissues more desperately needed elsewhere (examples include skin grafts, vein extraction for CABG, etc.). Sometimes an autograft is done to remove the tissue and then treat it or the person before returning it (examples include stem cell autograft and storing blood in advance of surgery). In a rotationplasty, a distal joint is used to replace a more proximal one; typically a foot or ankle joint is used to replace a knee joint. The person's foot is severed and reversed, the knee removed, and the tibia joined with the femur.

Allograft and Allotransplantation

An allograft is a transplant of an organ or tissue between two genetically non-identical members of the same species. Most human tissue and organ transplants are allografts. Due to the genetic difference between the organ and the recipient, the recipient's immune system will identify the organ as foreign and attempt to destroy it, causing transplant rejection. The risk of transplant rejection can be estimated by measuring the Panel reactive antibody level.

Isograft

A subset of allografts in which organs or tissues are transplanted from a donor to a genetically identical recipient (such as an identical twin). Isografts are differentiated from other types of transplants because while they are anatomically identical to allografts, they do not trigger an immune response.

Xenograft and Xenotransplantation

A transplant of organs or tissue from one species to another. An example is porcine heart valve transplant, which is quite common and successful. Another example is attempted piscine-primate (fish to non-human primate) transplant of islet (i.e. pancreatic or insular tissue) tissue. The latter research study was intended to pave the way for potential human use if successful. However, xenotransplantion is often an extremely dangerous type of transplant because of the increased risk of non-compatibility, rejection, and disease carried in the tissue.

Domino Transplants

In people with cystic fibrosis, where both lungs need to be replaced, it is a technically easier operation with a higher rate of success to replace both the heart and lungs of the recipient with those of the donor. As the recipient's original heart is usually healthy, it can then be transplanted into a second recipient in need of a heart transplant. Another example of this situation occurs with a special form of liver transplant in which the recipient suffers from familial amyloidotic polyneuropathy, a disease where the liver slowly produces a protein that damages other organs. The recipient's liver can then be transplanted into an older person for whom the effects of the disease will not necessarily contribute significantly to mortality.

This term also refers to a series of living donor transplants in which one donor donates to the highest recipient on the waiting list and the transplant center utilizes that donation to facilitate multiple transplants. These other transplants are otherwise impossible due to blood type or antibody barriers to transplantation. The "Good Samaritan" kidney is transplanted into one of the other recipients, whose donor in turn donates his or her kidney to an unrelated recipient. Depending on the person on the waiting list, this has sometimes been repeated for up to six pairs, with the final donor donating to the person at the top of the list. This method allows all organ recipients to get a transplant even if their living donor is not a match to them. This further benefits people below any of these recipients on waiting lists, as they move closer to the top of the list for a deceased-donor organ. Johns Hopkins Medical Center in Baltimore and Northwestern University's Northwestern Memorial Hospital have received significant attention for pioneering transplants of this kind.

In February 2012, the last link in a record 60-person domino chain of 30 kidney transplants was completed.

ABO-incompatible Transplants

Because very young children (generally under 12 months, but often as old as 24 months,) do not have a well-developed immune system, it is possible for them to receive organs from otherwise incompatible donors. This is known as ABO-incompatible (ABOi) transplantation. Graft survival and peoples mortality is approximately the same between ABOi and ABO-compatible (ABOc) recipients. While focus has been on infant heart transplants, the principles generally apply to other forms of solid organ transplantation.

The most important factors are that the recipient not have produced isohemagglutinins, and that they have low levels of T cell-independent antigens. United Network for Organ Sharing (UNOS) regulations allow for ABOi transplantation in children under two years of age if isohemagglutinin titers are 1:4 or below, and if there is no matching ABOc recipient. Studies have shown that the period under which a recipient may undergo ABOi transplantation may be prolonged by exposure to nonself A and B antigens. Furthermore, should the recipient (for example, type B-positive with a type AB-positive graft) require eventual retransplantation, the recipient may receive a new organ of either blood type.

Limited success has been achieved in ABO-incompatible heart transplants in adults, though this requires that the adult recipients have low levels of anti-A or anti-B antibodies. Kidney transplantation is more successful, with similar long-term graft survival rates to ABOc transplants.

Transplantation in Obese Individuals

Until recently, people labeled as obese were not considered appropriate candidates for renal transplantation. In 2009, the physicians at the University of Illinois Medical Center performed the first robotic kidney transplantation in an obese recipient and have continued to transplant people with Body Mass Index (BMI)'s over 35 using robotic surgery. As of January 2014, over 100 people that would otherwise be turned down because of their weight have successfully been transplanted.

Organs and Tissues Transplanted

Chest

- Heart (deceased-donor only)
- Lung (deceased-donor and living-related lung transplantation)
- Heart/Lung (deceased-donor and domino transplant)

Abdomen

- Kidney (deceased-donor and living-donor)
- Liver (deceased-donor and living-donor)
- Pancreas (deceased-donor only)
- Intestine (deceased-donor and living-donor)
- Stomach (deceased-donor only)
- Testis (deceased-donor and living-donor)

Tissues, Cells and Fluids

- Hand (deceased-donor only), the first recipient Clint Hallam
- Cornea (deceased-donor only) the ophthalmologist Eduard Zirm
- Skin, including face replant (autograft) and face transplant (extremely rare)
- Islets of Langerhans (pancreas islet cells) (deceased-donor and living-donor)
- Bone marrow/Adult stem cell (living-donor and autograft)
- Blood transfusion/Blood Parts Transfusion (living-donor and autograft)
- Blood Vessels (autograft and deceased-donor)
- Heart Valve (deceased-donor, living-donor and xenograft [porcine/bovine])
- Bone (deceased-donor and living-donor)

Types of Donor

Organ donors may be living or may have died of brain death or circulatory death. Most deceased donors are those who have been pronounced brain dead. Brain dead means the cessation of brain function, typically after receiving an injury (either traumatic or pathological) to the brain, or otherwise cutting off blood circulation to the brain (drowning, suffocation, etc.). Breathing is maintained via artificial sources, which, in turn, maintains heartbeat. Once brain death has been declared the person can be considered for organ donation. Criteria for brain death vary. Because less than 3% of all deaths in the U.S. are the result of brain death, the overwhelming majority of deaths are ineligible for organ donation, resulting in severe shortages.

Organ donation is possible after cardiac death in some situations, primarily when the person is severely brain injured and not expected to survive without artificial breathing and mechanical support. Independent of any decision to donate, a person's next-of-kin may decide to end artificial support. If the person is expected to expire within a short period of time after support is withdrawn, arrangements can be made to withdraw that support in an operating room to allow quick recovery of the organs after circulatory death has occurred.

Tissue may be recovered from donors who die of either brain or circulatory death. In general, tissues may be recovered from donors up to 24 hours past the cessation of heartbeat. In contrast to organs, most tissues (with the exception of corneas) can be preserved and stored for up to five years, meaning they can be "banked." Also, more than 60 grafts may be obtained from a single tissue donor. Because of these three factors—the ability to recover from a non-heart beating donor, the ability to bank tissue, and the number of grafts available from each donor—tissue transplants are much more common than organ transplants. The American Association of Tissue Banks estimates that more than one million tissue transplants take place in the United States each year.

Living Donor

In living donors, the donor remains alive and donates a renewable tissue, cell, or fluid (e.g., blood, skin), or donates an organ or part of an organ in which the remaining organ can regenerate or take on the workload of the rest of the organ (primarily single kidney donation, partial donation of liver, lung lobe, small bowel). Regenerative medicine may one day allow for laboratory-grown organs, using person's own cells via stem cells, or healthy cells extracted from the failing organs.

Deceased Donor

Deceased donors (formerly cadaveric) are people who have been declared brain-dead and whose organs are kept viable by ventilators or other mechanical mechanisms until they can be excised for transplantation. Apart from brain-stem dead donors, who have formed the majority of deceased donors for the last 20 years, there is increasing use of donation-after-circulatory-death-donors (formerly non-heart-beating donors) to increase the potential pool of donors as demand for transplants continues to grow. Prior to the recognition of brain death in the 1980s, all deceased organ donors had died of circulatory death. These organs have inferior outcomes to organs from a brain-dead donor; however, given the scarcity of suitable organs and the number of people who die waiting, any potentially suitable organ must be considered.

Allocation of Organs

In most countries there is a shortage of suitable organs for transplantation. Countries often have formal systems in place to manage the process of determining who is an organ donor and in what order organ recipients receive available organs.

The overwhelming majority of deceased-donor organs in the United States are allocated by federal contract to the Organ Procurement and Transplantation Network (OPTN), held since it was created by the Organ Transplant Act of 1984 by the United Network for Organ Sharing or UNOS. (UNOS does not handle donor cornea tissue; corneal donor tissue is usually handled by various eye banks.) Individual regional organ procurement organizations (OPOs), all members of the OPTN, are responsible for the identification of suitable donors and collection of the donated organs. UNOS then allocates organs based on the method considered most fair by the scientific leadership in the field. The allocation methodology varies somewhat by organ, and changes periodically. For example, liver allocation is based partially on MELD score (Model of End-Stage Liver Disease), an empirical score based on lab values indicative of the sickness of the person from liver disease. In 1984, the National Organ Transplant Act (NOTA) was passed which gave way to the Organ Procurement and Transplantation Network that maintains the organ registry and ensures equitable allocation of organs. The Scientific Registry of Transplant Recipients was also established to conduct ongoing studies into the evaluation and clinical status of organ transplants. In 2000 the Children's Health Act passed and required NOTA to consider special issues around pediatric patients and organ allocation (Services).

Experiencing somewhat increased popularity, but still very rare, is directed or targeted donation, in which the family of a deceased donor (often honoring the wishes of the deceased) requests an organ be given to a specific person. If medically suitable, the allocation system is subverted, and the organ is given to that person. In the United States, there are various lengths of waiting times due to the different availabilities of organs in different UNOS regions. In other countries such as the UK, only medical factors and the position on the waiting list can affect who receives the organ.

One of the more publicized cases of this type was the 1994 Chester and Patti Szuber transplant. This was the first time that a parent had received a heart donated by one of their own children. Although the decision to accept the heart from his recently killed child was not an easy decision, the Szuber family agreed that giving Patti's heart to her father would have been something that she would have wanted.

Access to organ transplantation is one reason for the growth of medical tourism.

Reasons for Donation and Ethical Issues

Living Related Donors

Living related donors donate to family members or friends in whom they have an emotional investment. The risk of surgery is offset by the psychological benefit of not losing someone related to them, or not seeing them suffer the ill effects of waiting on a list.

Paired Exchange

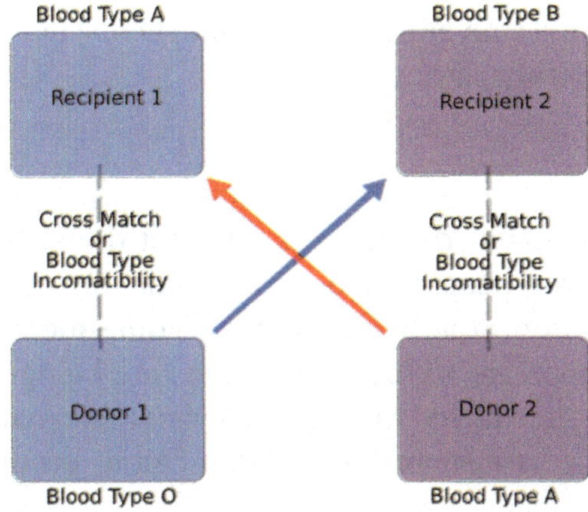

Diagram of an exchange between otherwise incompatible pairs

A "paired-exchange" is a technique of matching willing living donors to compatible recipients using serotyping. For example, a spouse may be willing to donate a kidney to their partner but cannot since there is not a biological match. The willing spouse's kidney is donated to a matching recipient who also has an incompatible but willing spouse. The second donor must match the first recipient to complete the pair exchange. Typically the surgeries are scheduled simultaneously in case one of the donors decides to back out and the couples are kept anonymous from each other until after the transplant.

Paired exchange programs were popularized in the New England Journal of Medicine article "Ethics of a paired-kidney-exchange program" in 1997 by L.F. Ross. It was also proposed by Felix T. Rapport in 1986 as part of his initial proposals for live-donor transplants "The case for a living emotionally related international kidney donor exchange registry" in *Transplant Proceedings*. A paired exchange is the simplest case of a much larger exchange registry program where willing donors are matched with any number of compatible recipients. Transplant exchange programs have been suggested as early as 1970: "A cooperative kidney typing and exchange program."

The first pair exchange transplant in the U.S. was in 2001 at Johns Hopkins Hospital. The first complex multihospital kidney exchange involving 12 people was performed in February 2009 by The Johns Hopkins Hospital, Barnes-Jewish Hospital in St. Louis and Integris Baptist Medical Center in Oklahoma City. Another 12-person multihospital kidney exchange was performed four weeks later by Saint Barnabas Medical Center in Livingston, New Jersey, Newark Beth Israel Medical Center and New York-Presbyterian Hospital. Surgical teams led by Johns Hopkins continue to pioneer in this field by having more complex chain of exchange such as eight-way multihospital kidney exchange. In December 2009, a 13 organ 13 recipient matched kidney exchange took place, coordinated through Georgetown University Hospital and Washington Hospital Center, Washington DC.

Paired-donor exchange, led by work in the New England Program for Kidney Exchange as well as at Johns Hopkins University and the Ohio OPOs may more efficiently allocate organs and lead to more transplants.

Good Samaritan

Good Samaritan or "altruistic" donation is giving a donation to someone not well-known to the donor. Some people choose to do this out of a need to donate. Some donate to the next person on the list; others use some method of choosing a recipient based on criteria important to them. Web sites are being developed that facilitate such donation. It has been featured in recent television journalism that over half of the members of the Jesus Christians, an Australian religious group, have donated kidneys in such a fashion.

Financial Compensation

Now monetary compensation for organ donors is being legalized in Australia, and strictly only in the case of kidney transplant in the case of Singapore (minimal reimbursement is offered in the case of other forms of organ harvesting by Singapore. Kidney disease organizations in both countries have expressed their support.

In compensated donation, donors get money or other compensation in exchange for their organs. This practice is common in some parts of the world, whether legal or not, and is one of the many factors driving medical tourism.

In the illegal black market the donors may not get sufficient after-operation care, the price of a kidney may be above $160,000, middlemen take most of the money, the operation is more dangerous to both the donor and receiver, and the receiver often gets hepatitis or HIV. In legal markets of Iran the price of a kidney is $2,000 to $4,000.

An article by Gary Becker and Julio Elias on "Introducing Incentives in the market for Live and Cadaveric Organ Donations" said that a free market could help solve the problem of a scarcity in organ transplants. Their economic modeling was able to estimate the price tag for human kidneys ($15,000) and human livers ($32,000).

In the United States, The National Organ Transplant Act of 1984 made organ sales illegal. In the United Kingdom, the Human Organ Transplants Act 1989 first made organ sales illegal, and has been superseded by the Human Tissue Act 2004.

In 2007, two major European conferences recommended against the sale of organs.

Recent development of web sites and personal advertisements for organs among listed candidates has raised the stakes when it comes to the selling of organs, and have also sparked significant ethical debates over directed donation, "good-Samaritan" donation, and the current U.S. organ allocation policy. Bioethicist Jacob M. Appel has argued that organ solicitation on billboards and the internet may actually increase the overall supply of organs.

Two books, *Kidney for Sale By Owner* by Mark Cherry (Georgetown University Press, 2005); and *Stakes and Kidneys: Why markets in human body parts are morally imperative* by James Stacey Taylor: (Ashgate Press, 2005); advocate using markets to increase the supply of organs available for transplantation. In a 2004 journal article Economist Alex Tabarrok argues that allowing organ sales, and elimination of organ donor lists will increase supply, lower costs and diminish social anxiety towards organ markets.

Iran has had a legal market for kidneys since 1988. The donor is paid approximately US$1200 by the government and also usually receives additional funds from either the recipient or local charities. *The Economist* and the Ayn Rand Institute approve and advocate a legal market elsewhere. They argued that if 0.06% of Americans between 19 and 65 were to sell one kidney, the national waiting list would disappear (which, the

Economist wrote, happened in Iran). The Economist argued that donating kidneys is no more risky than surrogate motherhood, which can be done legally for pay in most countries.

In Pakistan, 40 percent to 50 percent of the residents of some villages have only one kidney because they have sold the other for a transplant into a wealthy person, probably from another country, said Dr. Farhat Moazam of Pakistan, at a World Health Organization conference. Pakistani donors are offered $2,500 for a kidney but receive only about half of that because middlemen take so much. In Chennai, southern India, poor fishermen and their families sold kidneys after their livelihoods were destroyed by the Indian Ocean tsunami on 26 December 2004. About 100 people, mostly women, sold their kidneys for 40,000–60,000 rupees ($900–$1,350). Thilakavathy Agatheesh, 30, who sold a kidney in May 2005 for 40,000 rupees said, "I used to earn some money selling fish but now the post-surgery stomach cramps prevent me from going to work." Most kidney sellers say that selling their kidney was a mistake.

In Cyprus in 2010 police closed a fertility clinic under charges of trafficking in human eggs. The Petra Clinic, as it was known locally, imported women from Ukraine and Russia for egg harvesting and sold the genetic material to foreign fertility tourists. This sort of reproductive trafficking violates laws in the European Union. In 2010 Scott Carney reported for the Pulitzer Center on Crisis Reporting and the magazine Fast Company explored illicit fertility networks in Spain, the United States and Israel.

Forced Donation

There have been concerns that certain authorities are harvesting organs from people deem undesirable, such as prison populations. The World Medical Association stated that prisoners and other individuals in custody are not in a position to give consent freely, and therefore their organs must not be used for transplantation.

According to the Chinese Deputy Minister of Health, Huang Jiefu, approximately 95% of all organs used for transplantation are from executed prisoners. The lack of public organ donation program in China is used as a justification for this practice.

In July 2006, the Kilgour-Matas report stated, "the source of 41,500 transplants for the six year period 2000 to 2005 is unexplained" and "we believe that there has been and continues today to be large scale organ seizures from unwilling Falun Gong practitioners". Investigative journalist Ethan Gutmann estimates 65,000 Falun Gong practitioners were killed for their organs from 2000 to 2008. However 2016 reports updated the death toll of the 15-year period since the persecution of Falun Gong began putting the death toll at 150 thousand to 1.5 million.

In December 2006, after not getting assurances from the Chinese government about allegations relating to Chinese prisoners, the two major organ transplant hospitals in

Queensland, Australia stopped transplant training for Chinese surgeons and banned joint research programs into organ transplantation with China.

In May 2008, two United Nations Special Rapporteurs reiterated their requests for "the Chinese government to fully explain the allegation of taking vital organs from Falun Gong practitioners and the source of organs for the sudden increase in organ transplants that has been going on in China since the year 2000".

People in other parts of the world are responding to this availability of organs, and a number of individuals (including U.S. and Japanese citizens) have elected to travel to China or India as medical tourists to receive organ transplants which may have been sourced in what might be considered elsewhere to be unethical manner.

Usage

Some estimates of the number of transplants performed in various regions of the world have been derived from the Global Burden of Disease Study.

Fig. 1. **Distribution of solid organ transplantation activity, by region used in the Global Burden of Disease Study, 2006–2011**

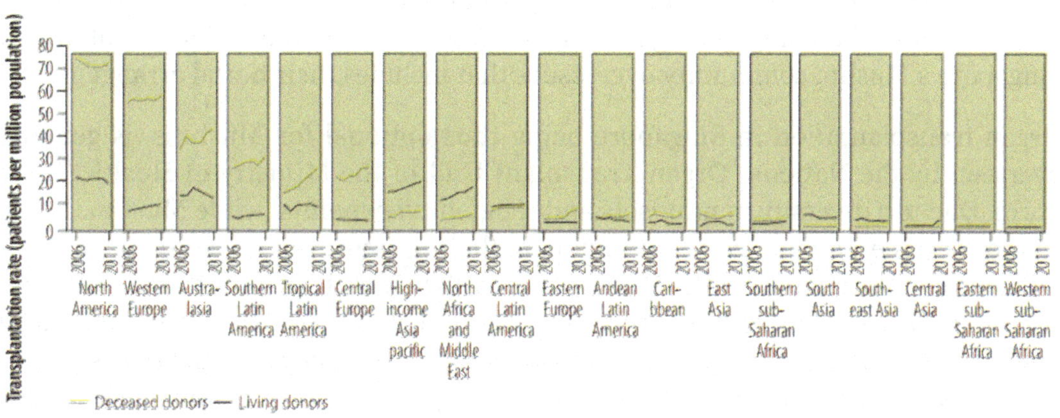

Transplantation of organs in different regions in 2000

	Kidney (pmp*)	Liver (pmp)	Heart (pmp)
United States	52	19	8
Europe	27	10	4
Africa	11	3.5	1
Asia	3	0.3	0.03
Latin America	13	1.6	0.5

*All numbers per million population

According to the Council of Europe, Spain through the Spanish Transplant Organization shows the highest worldwide rate of 35.1 donors per million population in 2005 and 33.8 in 2006. In 2011, it was 35.3.

In addition to the citizens waiting for organ transplants in the U.S. and other developed nations, there are long waiting lists in the rest of the world. More than 2 million people need organ transplants in China, 50,000 waiting in Latin America (90% of which are waiting for kidneys), as well as thousands more in the less documented continent of Africa. Donor bases vary in developing nations.

In Latin America the donor rate is 40–100 per million per year, similar to that of developed countries. However, in Uruguay, Cuba, and Chile, 90% of organ transplants came from cadaveric donors. Cadaveric donors represent 35% of donors in Saudi Arabia. There is continuous effort to increase the utilization of cadaveric donors in Asia, however the popularity of living, single kidney donors in India yields India a cadaveric donor prevalence of less than 1 pmp.

Traditionally, Muslims believe body desecration in life or death to be forbidden, and thus many reject organ transplant. However most Muslim authorities nowadays accept the practice if another life will be saved. As an example, it may be assumed in countries such as Singapore with a cosmopolitan populace that includes Muslims, a special Majlis Ugama Islam Singapura governing body is formed to look after the interests of Singapore's Muslim community over issues that includes their burial arrangements.

Organ transplantation in Singapore being thus optional for Muslims, is generally overseen by the National Organ Transplant Unit of the Ministry of Health (Singapore). Due to a diversity in mindsets and religious viewpoints, while Muslims on this island are generally not expected to donate their organs even upon death, youths in Singapore are educated on the Human Organ Transplant Act at the age of 18 which is around the age of military conscription. The Organ Donor Registry maintains two types of information, firstly people of Singapore that donate their organs or bodies for transplantation, research or education upon their death, under the Medical (Therapy, Education and Research) Act (MTERA), and secondly people that object to the removal of kidneys, liver, heart and corneas upon death for the purpose of transplantation, under the Human Organ Transplant Act (HOTA). The Live On social awareness movement is also formed to educate Singaporeans on organ donation.

Organ transplantation in China has taken place since the 1960s, and China has one of the largest transplant programmes in the world, peaking at over 13,000 transplants a year by 2004. Organ donation, however, is against Chinese tradition and culture, and involuntary organ donation is illegal under Chinese law. China's transplant programme attracted the attention of international news media in the 1990s due to ethical concerns about the organs and tissue removed from the corpses of executed criminals being commercially traded for transplants. In 2006 it became clear that about 41,500 organs had been sourced from Falun Gong practitioners in China since 2000. With regard to organ transplantation in Israel, there is a severe organ shortage due to religious objections by some rabbis who oppose all organ donations and others who advocate that a rabbi participate in all decision making regarding a particular donor. One third of all heart transplants performed on Israelis are done in the Peoples' Republic of China; others are done in Europe. Dr. Jacob Lavee, head of the heart-transplant unit, Sheba Medical Center, Tel Aviv, believes that "transplant tourism" is unethical and Israeli insurers should not pay for it. The organization HODS (Halachic Organ Donor Society) is working to increase knowledge and participation in organ donation among Jews throughout the world.

Transplantation rates also differ based on race, sex, and income. A study done with people beginning long term dialysis showed that the sociodemographic barriers to renal transplantation present themselves even before patients are on the transplant list. For example, different groups express definite interest and complete pretransplant workup at different rates. Previous efforts to create fair transplantation policies had focused on people currently on the transplantation waiting list.

History

Successful human allotransplants have a relatively long history of operative skills that were present long before the necessities for post-operative survival were discovered. Rejection and the side effects of preventing rejection (especially infection and nephropathy) were, are, and may always be the key problem.

Several apocryphal accounts of transplants exist well prior to the scientific understanding and advancements that would be necessary for them to have actually occurred. The Chinese physician Pien Chi'ao reportedly exchanged hearts between a man of strong spirit but weak will with one of a man of weak spirit but strong will in an attempt to achieve balance in each man. Roman Catholic accounts report the 3rd-century saints Damian and Cosmas as replacing the gangrenous or cancerous leg of the Roman deacon Justinian with the leg of a recently deceased Ethiopian. Most accounts have the saints performing the transplant in the 4th century, many decades after their deaths; some accounts have them only instructing living surgeons who performed the procedure.

The more likely accounts of early transplants deal with skin transplantation. The first reasonable account is of the Indian surgeon Sushruta in the 2nd century BC, who used autografted skin transplantation in nose reconstruction, a rhinoplasty. Success or failure of these procedures is not well documented. Centuries later, the Italian surgeon Gasparo Tagliacozzi performed successful skin autografts; he also failed consistently with allografts, offering the first suggestion of rejection centuries before that mechanism could possibly be understood. He attributed it to the "force and power of individuality" in his 1596 work *De Curtorum Chirurgia per Insitionem*.

Alexis Carrel: 1912's Nobel Prize for his work on organ transplantation.

The first successful corneal allograft transplant was performed in 1837 in a gazelle model; the first successful human corneal transplant, a keratoplastic operation, was performed by Eduard Zirm at Olomouc Eye Clinic, now Czech Republic, in 1905. The first transplant in the modern sense – the implantation of organ tissue in order to replace an organ function – was a thyroid transplant in 1883. It was performed by the Swiss surgeon and later Nobel laureate Theodor Kocher. In the preceding decades Kocher had perfected the removal of excess thyroid tissue in cases of goiter to an extent that he was able to remove the whole organ without the person dying from the operation. Kocher carried out the total removal of the organ in some cases as a measure to prevent recurrent goiter. By 1883, the surgeon noticed that the complete removal of the organ leads to a complex of particular symptoms that we today have learned to associate with a lack of thyroid hormone. Kocher reversed these symptoms by implanting thyroid tissue to these people and thus performed the first organ transplant. In the following years Kocher and other surgeons used thyroid transplantation also to treat thyroid deficiency that appeared spontaneously, without a preceding organ removal. Thyroid transplantation became the model for a whole new therapeutic strategy: organ transplantation. After the example of the thyroid, other organs were transplanted in the decades around 1900. Some of these transplants were done in animals for purposes of research, where organ removal and transplantation became a successful strategy of investigating the function of organs. Kocher was awarded his Nobel Prize in 1909 for the discovery of the function of the thyroid gland. At the same time, organs were also transplanted for treating diseases in humans. The thyroid gland became the model for transplants of adrenal and parathyroid glands, pancreas, ovary, testicles and kidney. By 1900, the idea that one can successfully treat internal diseases by replacing a failed organ through transplantation had been generally accepted. Pioneering work in the surgical technique of transplantation was made in the early 1900s by the French surgeon Alexis Carrel, with Charles Guthrie, with the transplantation of arteries or veins. Their skillful anastomosis operations and the new suturing techniques laid the groundwork for later transplant surgery and won Carrel the 1912 Nobel Prize in Physiology or Medicine. From 1902, Carrel performed transplant experiments on dogs. Surgically successful in moving kidneys, hearts, and spleens, he was one of the first to identify the problem of rejection, which remained insurmountable for decades. The discovery of transplant immunity by the German surgeon Georg Schöne, various strategies of matching donor and recipient, and the use of different agents for immune suppression did not result in substantial improvement so that organ transplantation was largely abandoned after WWI.

Major steps in skin transplantation occurred during the First World War, notably in the work of Harold Gillies at Aldershot. Among his advances was the tubed pedicle graft, which maintained a flesh connection from the donor site until the graft established its own blood flow. Gillies' assistant, Archibald McIndoe, carried on the work into the Second World War as reconstructive surgery. In 1962, the first successful replantation surgery was performed – re-attaching a severed limb and restoring (limited) function and feeling.

Transplant of a single gonad (testis) from a living donor was carried out in early July 1926 in Zaječar, Serbia, by a Russian emigré surgeon Dr. Peter Vasil'evič Kolesnikov. The donor was a convicted murderer, one Ilija Krajan, whose death sentence was commuted to 20 years imprisonment, and he was led to believe that it was done because he had donated his testis to an elderly medical doctor. Both the donor and the receiver survived, but charges were brought in a court of law by the public prosecutor against Dr. Kolesnikov, not for performing the operation, but for lying to the donor.

The first attempted human deceased-donor transplant was performed by the Ukrainian surgeon Yuri Voronoy in the 1930s; but failed due to Ischemia. Joseph Murray and J. Hartwell Harrison performed the first successful transplant, a kidney transplant between identical twins, in 1954, because no immunosuppression was necessary for genetically identical individuals.

In the late 1940s Peter Medawar, working for the National Institute for Medical Research, improved the understanding of rejection. Identifying the immune reactions in 1951, Medawar suggested that immunosuppressive drugs could be used. Cortisone had been recently discovered and the more effective azathioprine was identified in 1959, but it was not until the discovery of cyclosporine in 1970 that transplant surgery found a sufficiently powerful immunosuppressive.

Joseph Murray's success with the kidney led to attempts with other organs. There was a successful deceased-donor lung transplant into a lung cancer sufferer in June 1963 by James Hardy in Jackson, Mississippi. The person survived for eighteen days before dying of kidney failure. Thomas Starzl of Denver attempted a liver transplant in the same year, but he was not successful until 1967.

The heart was a major prize for transplant surgeons. But over and above rejection issues, the heart deteriorates within minutes of death, so any operation would have to be performed at great speed. The development of the heart-lung machine was also needed. Lung pioneer James Hardy attempted a human heart transplant in 1964, but when a premature failure of the recipient's heart caught Hardy with no human donor, he used a chimpanzee heart, which failed very quickly. The first success was achieved on 3 December 1967, by Christiaan Barnard in Cape Town, South Africa. Louis Washkansky, the recipient, survived for eighteen days amid what many saw as a distasteful publicity circus. The media interest prompted a spate of heart transplants. Over a hundred were performed in 1968–1969, but almost all the people died within 60 days. Barnard's second patient, Philip Blaiberg, lived for 19 months.

It was the advent of cyclosporine that altered transplants from research surgery to life-saving treatment. In 1968 surgical pioneer Denton Cooley performed 17 transplants, including the first heart-lung transplant. Fourteen of his patients were dead within six months. By 1984 two-thirds of all heart transplant patients survived for five years or more. With organ transplants becoming commonplace, limited only by donors, surgeons moved on to riskier fields, including multiple-organ transplants on humans and whole-body transplant research on animals. On 9 March 1981, the first successful heart-lung transplant took place at Stanford University Hospital. The head surgeon, Bruce Reitz, credited the patient's recovery to cyclosporine-A.

As the rising success rate of transplants and modern immunosuppression make transplants more common, the need for more organs has become critical. Transplants from living donors, especially relatives, have become increasingly common. Additionally, there is substantive research into xenotransplantation, or transgenic organs; although these forms of transplant are not yet being used in humans, clinical trials involving the use of specific cell types have been conducted with promising results, such as using porcine islets of Langerhans to treat type 1 diabetes. However, there are still many problems that would need to be solved before they would be feasible options in people requiring transplants.

Recently, researchers have been looking into means of reducing the general burden of immunosuppression. Common approaches include avoidance of steroids, reduced exposure to calcineurin inhibitors, and other means of weaning drugs based on patient outcome and function. While short-term outcomes appear promising, long-term outcomes are still unknown, and in general, reduced immunosuppression increases the risk of rejection and decreases the risk of infection. The risk of early rejection is increased if corticosteroid immunosuppression are avoided or withdrawn after kidney transplantation.

Many other new drugs are under development for transplantation. The emerging field of regenerative medicine promises to solve the problem of organ transplant rejection by regrowing organs in the lab, using person's own cells (stem cells or healthy cells extracted from the donor site).

Timeline of Successful Transplants

- 1823: First skin autograft-transplantation of skin tissue from one location on an individual's body to another location (Germany) In 1823 Carl Bunger, a German surgeon documents the first modern successful skin graft on a person. Bunger was repairing a person with a nose also destroyed by syphilis. He grafted a small chunk of full thickness flesh from the inner thigh to the nose successfully, in a method very reminiscent of Sushrutha's.
- 1905: First successful cornea transplant by Eduard Zirm (Czech Republic)
- 1908: First skin allograft-transplantation of skin from a donor to a recipient (Switzerland)
- 1933: First successful cadaveric AB-0 incompatible kidney transplant (donor was B(III) and the recipient has 0(I)) by Yuriu Yu. Voronoy (USSR)
- 1950: First successful kidney transplant by Dr. Richard H. Lawler (Chicago, U.S.A.)
- 1954: First living related kidney transplant (identical twins) (U.S.A.)
- 1954: Brazil's first successful corneal transplant, the first liver (Brazil)
- 1955: First heart valve allograft into descending aorta (Canada)
- 1962: First kidney transplant from a deceased donor (U.S.A.)
- 1965: Australia's first successful (living) kidney transplant (Queen Elizabeth Hospital, SA, Australia)
- 1966: First successful pancreas transplant by Richard Lillehei and William Kelly (Minnesota, U.S.A.)
- 1967: First successful liver transplant by Thomas Starzl (Denver, U.S.A.)
- 1967: First successful heart transplant by Christian Barnard (Cape Town, South Africa)
- 1981: First successful heart/lung transplant by Bruce Reitz (Stanford, U.S.A.)
- 1983: First successful lung lobe transplant by Joel Cooper at the Toronto General Hospital (Toronto, Canada)
- 1984: First successful double organ transplant by Thomas Starzl and Henry T. Bahnson (Pittsburgh, U.S.A.)
- 1986: First successful double-lung transplant (Ann Harrison) by Joel Cooper at the Toronto General Hospital (Toronto, Canada)

- 1995: First successful laparoscopic live-donor nephrectomy by Lloyd Ratner and Louis Kavoussi (Baltimore, U.S.A.)
- 1997: First successful allogeneic vascularized transplantation of a fresh and perfused human knee joint by Gunther O. Hofmann
- 1997: Illinois' first living donor kidney-pancreas transplant and first robotic living donor pancreatectomy in the U.S.A. University of Illinois Medical Center
- 1998: First successful live-donor partial pancreas transplant by David Sutherland (Minnesota, U.S.A.)
- 1998: First successful hand transplant by Dr. Jean-Michel Dubernard (Lyon, France)
- 1998: United States' first adult-to-adult living donor liver transplant University of Illinois Medical Center
- 1999: First successful tissue engineered bladder transplanted by Anthony Atala (Boston Children's Hospital, U.S.A.)
- 2000: First robotic donor nephrectomy for a living-donor kidney transplant in the world University of Illinois Medical Center
- 2004: First liver and small bowel transplants from same living donor into same recipient in the world University of Illinois Medical Center
- 2005: First successful ovarian transplant by Dr. P. N. Mhatre (Wadia Hospital, Mumbai, India)
- 2005: First successful partial face transplant (France)
- 2005: First robotic hepatectomy in the United States University of Illinois Medical Center
- 2006: Illinois' first paired donation for ABO incompatible kidney transplant University of Illinois Medical Center
- 2006: First jaw transplant to combine donor jaw with bone marrow from the patient, by Eric M. Genden (Mount Sinai Hospital, New York City, U.S.A.)
- 2006: First successful human penis transplant (later reversed after 15 days due to 44-year-old recipient's wife's psychological rejection) (Guangzhou, China)
- 2008: First successful complete full double arm transplant by Edgar Biemer, Christoph Höhnke and Manfred Stangl (Technical University of Munich, Germany)
- 2008: First baby born from transplanted ovary. The transplant was carried out

by Dr Sherman Silber at the Infertility Centre of St Louis in Missouri. The donor is her twin sister.

- 2008: First transplant of a human windpipe using a patient's own stem cells, by Paolo Macchiarini (Barcelona, Spain)
- 2008: First successful transplantation of near total area (80%) of face, (including palate, nose, cheeks, and eyelid) by Maria Siemionow (Cleveland Clinic, U.S.A.)
- 2009: Worlds' first robotic kidney transplant in an obese patient University of Illinois Medical Center
- 2010: First full facial transplant by Dr. Joan Pere Barret and team (Hospital Universitari Vall d'Hebron on 26 July 2010, in Barcelona, Spain)
- 2011: First double leg transplant by Dr. Cavadas and team (Valencia's Hospital, La Fe, Spain)
- 2012: First Robotic Alloparathyroid transplant. University of Illinois Chicago
- 2013: First successful entire face transplantation as an urgent life-saving surgery at Maria Skłodowska-Curie Institute of Oncology branch in Gliwice, Poland.
- 2014: First successful uterine transplant resulting in live birth (Sweden)
- 2014: First successful penis transplant. (South Africa)
- 2014: First neonatal organ transplant. (U.K.)

Society and Culture

Comparative Costs

One of the driving forces for illegal organ trafficking and for "transplantation tourism" is the price differences for organs and transplant surgeries in different areas of the world. According to the New England Journal of Medicine, a human kidney can be purchased in Manila for $1000–$2000, but in urban Latin America a kidney may cost more than $10,000. Kidneys in South Africa have sold for as high as $20,000. Price disparities based on donor race are a driving force of attractive organ sales in South Africa, as well as in other parts of the world.

In China, a kidney transplant operation runs for around $70,000, liver for $160,000, and heart for $120,000. Although these prices are still unattainable to the poor, compared to the fees of the United States, where a kidney transplant may demand $100,000, a liver $250,000, and a heart $860,000, Chinese prices have made China a major provider of organs and transplantation surgeries to other countries.

In India, a kidney transplant operation runs for around as low as $5000.

Safety

In the United States of America, tissue transplants are regulated by the U.S. Food and Drug Administration (FDA) which sets strict regulations on the safety of the transplants, primarily aimed at the prevention of the spread of communicable disease. Regulations include criteria for donor screening and testing as well as strict regulations on the processing and distribution of tissue grafts. Organ transplants are not regulated by the FDA.

In November 2007, the CDC reported the first-ever case of HIV and Hepatitis C being simultaneously transferred through an organ transplant. The donor was a 38-year-old male, considered "high-risk" by donation organizations, and his organs transmitted HIV and Hepatitis C to four organ recipients. Experts say that the reason the diseases did not show up on screening tests is probably because they were contracted within three weeks before the donor's death, so antibodies would not have existed in high enough numbers to detect. The crisis has caused many to call for more sensitive screening tests, which could pick up antibodies sooner. Currently, the screens cannot pick up on the small number of antibodies produced in HIV infections within the last 90 days or Hepatitis C infections within the last 18–21 days before a donation is made.

NAT (nucleic acid testing) is now being done by many organ procurement organizations and is able to detect HIV and Hepatitis C directly within seven to ten days of exposure to the virus.

Transplant Laws

Both developing and developed countries have forged various policies to try to increase the safety and availability of organ transplants to their citizens. Austria, Brazil, France, Italy, Poland and Spain have ruled all adults potential donors with the "opting out" policy, unless they attain cards specifying not to be. However, whilst potential recipients in developing countries may mirror their more developed counterparts in desperation, potential donors in developing countries do not. The Indian government has had difficulty tracking the flourishing organ black market in their country, but in recent times it has amended its organ transplant law to make punishment more stringent for commercial dealings in organs. It has also included new clauses in the law to support deceased organ donation, such as making it mandatory to request for organ donation in case of brain death. Other countries victimized by illegal organ trade have also implemented legislative reactions. Moldova has made international adoption illegal in fear of organ traffickers. China has made selling of organs illegal as of July 2006 and claims that all prisoner organ donors have filed consent. However, doctors in other countries, such as the United Kingdom, have accused China of abusing its high capital punishment rate. Despite these efforts, illegal organ trafficking continues to thrive and can be attributed to corruption in healthcare systems, which has been traced as high up as the doctors themselves in China and Ukraine, and the blind eye economically strained governments and health care programs must sometimes turn to organ trafficking. Some organs are also shipped to Uganda and the Netherlands. This was a main product in the triangular trade in 1934.

Starting on 1 May 2007, doctors involved in commercial trade of organs will face fines and suspensions in China. Only a few certified hospitals will be allowed to perform organ transplants in order to curb illegal transplants. Harvesting organs without donor's consent was also deemed a crime.

On 27 June 2008, Indonesian, Sulaiman Damanik, 26, pleaded guilty in Singapore court for sale of his kidney to CK Tang's executive chair, Tang Wee Sung, 55, for 150 million rupiah (S$22,200). The Transplant Ethics Committee must approve living donor kidney transplants. Organ trading is banned in Singapore and in many other countries to prevent the exploitation of "poor and socially disadvantaged donors who are unable to make informed choices and suffer potential medical risks." Toni, 27, the other accused, donated a kidney to an Indonesian patient in March, alleging he was the patient's adopted son, and was paid 186 million rupiah (20,200USD). Upon sentence, both would suffer each, 12 months in jail or 10,000 Singapore dollars (7,600 USD) fine.

In an article appearing in the April 2004 issue of Econ Journal Watch, economist Alex Tabarrok examined the impact of direct consent laws on transplant organ availability. Tabarrok found that social pressures resisting the use of transplant organs decreased over time as the opportunity of individual decisions increased. Tabarrok concluded his study suggesting that gradual elimination of organ donation restrictions and move to a free market in organ sales will increase supply of organs and encourage broader social acceptance of organ donation as a practice.

Ethical Concerns

The existence and distribution of organ transplantation procedures in developing countries, while almost always beneficial to those receiving them, raise many ethical concerns. Both the source and method of obtaining the organ to transplant are major ethical issues to consider, as well as the notion of distributive justice. The World Health Organization argues that transplantations promote health, but the notion of "transplantation tourism" has the potential to violate human rights or exploit the poor, to have unintended health consequences, and to provide unequal access to services, all of which ultimately may cause harm. Regardless of the "gift of life", in the context of developing countries, this might be coercive. The practice of coercion could be considered exploitative of the poor population, violating basic human rights according to Articles 3 and 4 of the Universal Declaration of Human Rights. There is also a powerful opposing view, that trade in organs, if properly and effectively regulated to ensure that the seller is fully informed of all the consequences of donation, is a mutually beneficial transaction between two consenting adults, and that prohibiting it would itself be a violation of Articles 3 and 29 of the Universal Declaration of Human Rights.

Even within developed countries there is concern that enthusiasm for increasing the supply of organs may trample on respect for the right to life. The question is made even more complicated by the fact that the "irreversibility" criterion for legal death cannot be adequately defined and can easily change with changing technology.

Artificial Organ Transplantation

Surgeons, notably Paolo Macchiarini, in Sweden performed the first implantation of a synthetic trachea in July 2011, for a 36-year-old patient who was suffering from cancer. Stem cells taken from the patient's hip were treated with growth factors and incubated on a plastic replica of his natural trachea.

According to information uncovered by the Swedish documentary "Dokument Inifrån: Experimenten" (Swedish: "Documents from the Inside: The Experiments") the patient, Andemariam went on to suffer an increasingly terrible and eventually bloody cough to dying, incubated, in the hospital. At that point, determined by autopsy, 90% of the synthetic windpipe had come loose. He allegedly made several trips to Macchiarini for his complications, and at one point had surgery again to have his synthetic windpipe replaced, but Macchiarini was notoriously difficult to get an appointment with. According to the autopsy, the old synthetic windpipe did not appear to have been replaced.

Macchiarini's academic credentials have been called into question and he has recently been accused of alleged research misconduct.

Research

An early-stage medical laboratory and research company, called Organovo, designs and develops functional, three dimensional human tissue for medical research and therapeutic applications. The company utilizes its NovoGen MMX Bioprinter for 3D bioprinting. Organovo anticipates that the bioprinting of human tissues will accelerate the preclinical drug testing and discovery process, enabling treatments to be created more quickly and at lower cost. Additionally, Organovo has long-term expectations that this technology could be suitable for surgical therapy and transplantation.

Sperm Bank

A sperm bank, semen bank or cryobank is a facility or enterprise that collects and stores human sperm from sperm donors for use by women who need donor-provided sperm to achieve pregnancy. Sperm donated by the sperm donor is known as donor sperm, and the process for introducing the sperm into the woman is called artificial insemination, which is a form of third-party reproduction.

From a medical perspective, a pregnancy achieved using donor sperm is no different from a pregnancy achieved using partner sperm, and it is also no different from a pregnancy achieved by sexual intercourse.

General

Sperm banks enable greater control, especially in relation to the access and timing of pregnancies, since they check and screen potential donors, and provide formerly infertile couples or single women the chance to have babies.

Some controversy stems from the fact that donors father children for others and usually take no part in the upbringing of such children. In addition, some sperm banks supply or treat single women and coupled lesbians so that they can have their own biological children by a donor. Donors may not have a say in who may use their sperm.

The increasing range of services which is available through sperm banks nevertheless enables more couples to have choices over the whole issue of reproduction. Women may choose to use an anonymous donor who will not be a part of family life, or they may choose known donors who may be contacted later in life by the donor children. Women may choose to use a surrogate to bear their children, using eggs provided by the woman and sperm from a donor. Sperm banks often provide services which enable a woman to have subsequent pregnancies by the same donor, but equally, women may choose to have children by a number of different donors. Sperm banks sometimes enable a woman to choose the sex of her child, enabling even greater control over the way families are planned. Sperm banks increasingly adopt a less formal approach to the provision of their services thereby enabling people to take a relaxed approach to their own individual requirements.

Men who choose to donate sperm through a sperm bank also have the security of knowing that they are helping women or childless couples to have children in circumstances where they, as the biological father, will not have any legal or other responsibility for the children produced from their sperm. Whether a donor is anonymous or not, this factor is important in allowing sperm banks to recruit sperm donors and to use their sperm to produce whatever number of pregnancies from each donor as are permitted where they operate, or alternatively, whatever number they decide.

However, in many parts of the world sperm banks are not allowed to be established or to operate. Sperm banks do not provide a cure for infertility in that it is the sperm donor who reproduces himself, not a partner of the recipient woman. Most societies are built upon the family model and sperm banks may be seen as a threat to this, particularly where a sperm bank makes its services available to unmarried women.

Where sperm banks are allowed to operate they are often controlled by local legislation which is primarily intended to protect the unborn child, but which may also provide a compromise between the conflicting views which surround their operation. A particular example of this is the control which is often placed on the number of children which a single donor may father and which may be designed to protect against consanguinity. However, such legislation usually cannot prevent a sperm bank from supplying donor sperm outside the jurisdiction in which it operates, and neither can it prevent sperm donors from donating elsewhere during their lives. There is an acute shortage of sperm donors in many parts of the world and there is obvious pressure from many quarters for donor sperm from those willing and able to provide it to be made available as safely and as freely as possible.

Recruitment

The finding of a potential sperm donor and motivating him to actually donate sperm is typically called recruitment. A sperm bank recruits donors by advertising, often in colleges and in local newspapers, and also on the internet.

A donor must be a fit healthy male, normally between 18 and 45 years of age, who is willing to undergo frequent and rigorous testing and who is willing to donate his sperm so that it can be used to impregnate women who are unrelated to and unknown by him. The donor must agree to relinquish all legal rights to all children which result from his donations. The donor must produce his sperm at the sperm bank thus enabling the identity of the donor, once proven, always to be ascertained, and also enabling fresh samples of sperm to be produced for immediate processing.

Screening of Donors

A sperm donor must generally meet specific requirements regarding age and medical history.

Sperm banks typically screen potential donors for a range of diseases and disorders, including genetic diseases, chromosomal abnormalities and sexually transmitted infections that may be transmitted through sperm. The screening procedure generally also includes a quarantine period, in which the samples are frozen and stored for at least 6 months after which the donor will be re-tested for the STIs. This is to ensure no new infections have been acquired or have developed during the period of donation. Providing the result is negative, the sperm samples can be released from quarantine and used in treatments. Children conceived through sperm donation have a birth defect rate of almost a fifth compared with the general population.

A sperm bank takes a number of steps to ensure the health and quality of the sperm which it supplies and it will inform customers of the checks which it undertakes, providing relevant information about individual donors. A sperm bank will usually guarantee the quality and number of motile sperm available in a sample after thawing. They will try to select men as donors who are particularly fertile and whose sperm will survive the freezing and thawing process. Samples are often sold as containing a particular number of motile sperm per millilitre, and different types of sample may be sold by a sperm bank for differing types of use, e.g. ICI or IUI.

The sperm will be checked to ensure its fecundity and also to ensure that motile sperm will survive the freezing process. If a man is accepted onto the sperm bank's program as a sperm donor, his sperm will be constantly monitored, the donor will be regularly checked for infectious diseases, and samples of his blood will be taken at regular intervals. A sperm bank may provide a donor with dietary supplements containing herbal or mineral substances such as maca, zinc, vitamin E and arginine which are designed to improve the quality and quantity of the donor's semen, as well as reducing the refractory time (i.e. the time between viable ejaculations). All sperm is frozen in straws or vials and stored for a minimum of 6 months before being released for sale and use to ensure that the donor remains healthy.

Donors are subject to tests for infectious diseases such as human immunoviruses HIV (HIV-1 and HIV-2), human T-cell lymphotropic viruses (HTLV-1 and HTLV-2), syphilis, chlamydia, gonorrhea, Hepatitis B virus, Hepatitis C virus, cytomegalovirus (CMV), Trypanosoma cruzi and Malaria as well as hereditary diseases such as cystic fibrosis, Sickle cell anemia, Familial Mediterranean fever, Gaucher's disease, Thalassaemia, Tay-Sachs disease, Canavan's disease, Familial dysautonomia, Congenital adrenal hyperplasia Carnitine transporter deficiency and Karyotyping 46XY. Karyotyping is not a requirement in either EU or the US but some sperm banks choose to test donors as an extra service to the customer.

A sperm donor may also be required to produce his medical records and those of his family, often for several generations. A sperm sample is usually tested micro-biologically at the sperm bank before it is prepared for freezing and subsequent use. A sperm donor's blood group may also be registered to ensure compatibility with the recipient.

Sexually active gay men are prohibited or discouraged from donating in some countries, including the United States. Some sperm banks also screen out some potential donors based on height, baldness, and family medical history.

Donor payment

The majority of sperm donors who donate their sperm through a sperm bank receive some kind of payment, although this is rarely a significant amount. A review including 29 studies from 9 countries came to the result that the amount of money actual donors received for their donation varied from $10 to €70 per donation or sample. The payments vary from the situation in the United Kingdom where donors are only entitled to their expenses in connection with the donation, to the situation with some US sperm banks where a donor receives a set fee for each donation plus an additional amount for each vial stored. At one prominent California sperm bank for example, TSBC, donors receive roughly $50 for each donation (ejaculation) which has acceptable motility/survival rates both at donation and at a test-thaw a couple of days later. Because of the requirement for the two-day celibacy period before donation, and geographical factors which usually require the donor to travel, it is not a viable way to earn a significant income—and is far less lucrative than selling human eggs. Some private donors may seek remuneration although others donate for altruistic reasons. According to the EU Tissue Directive donors in EU may only receive compensation, which is strictly limited to making good the expenses and inconveniences related to the donation. A survey among sperm donors in Cryos International Sperm bank showed that altruistic as well as financial motives were the main factors for becoming a donor. However, when the compensation was increased 100% in 2004 (to DKK 500) it had no significant impact on neither the numbers of new donor candidates coming in nor the frequency of donations from the existing donors. When the compensation was reduced to the previous level (DKK 250) again one year later in 2005 there was no effect either. This led to the assumption that altruism is the main motive and that financial compensation is secondary.

Collection

A sperm donor will usually donate sperm to a sperm bank under a contract, which would typically specify the period during which the donor will be required to produce sperm, which generally ranges from 6–24 months depending on the number of pregnancies which the sperm bank intends to produce from the donor. Where local regulations or the sperm bank's own rules limit the number of pregnancies which a single donor can achieve, his donations will be limited for this reason. In the United Kingdom, for example, where a donor is not permitted to father more than ten families, a sperm bank will generally need a maximum of 100 straws prepared for IUI insemination, so that a man will generally not donate for more than twelve months.

However, not all donors complete the intended program of donations. If a sperm bank has access to world markets e.g. by direct sales, or sales to clinics outside their own jurisdiction, a man may donate for a longer period than two years, as the risk of consanguinity is reduced (although local laws vary widely). Some sperm banks with access to world markets impose their own rules on the number of pregnancies which can be achieved in a given regional area or a state or country, and these sperm banks may permit donors to donate for four or five years, or even longer. Faced with a growing demand for donor sperm, sperm banks may try to maximise the use of a donor whilst still reducing the risk of consanguinity.

The contract may also specify the place and hours for donation, a requirement to notify the sperm bank in the case of acquiring a sexual infection, and the requirement not to have intercourse or to masturbate for a period of usually 2–3 days before making a donation.

A sperm donor generally produces and collects sperm at a sperm bank or clinic by masturbation in a private room or cabin, known as a 'men's production room' (UK), 'donor cabin' (DK) or a masturbatorium (USA). Many of these facilities contain pornography such as videos/DVD, magazines, and/or photographs which may assist the donor in becoming aroused in order to facilitate production of the ejaculate, also known as the "semen sample". In some circumstances, it may also be possible for semen from donors to be collected during sexual intercourse with the use of a collection condom.

Processing Sperm

Sperm banks and clinics usually 'wash' the sperm sample to extract sperm from the rest of the material in the semen. A cryoprotectant semen extender is added if the sperm is to be placed in frozen storage. One sample can produce 1-20 vials or straws, depending on the quantity of the ejaculate and whether the sample is 'washed' or 'unwashed'. 'Unwashed' samples are used for intracervical insemination (ICI) treatments, and 'washed' samples are used in intrauterine insemination (IUI) and for in-vitro fertilization (IVF) procedures.

Storage

The sperm is stored in small vials or straws holding between 0.4 and 1.0 ml of sperm and cryogenically preserved in liquid nitrogen tanks. It has been proposed that there should be an upper limit on how long frozen sperm can be stored, however, a baby has been conceived in the United Kingdom using sperm frozen for 21 years.

Before freezing, sperm may be prepared (washed or left unwashed) so that it can be used for intracervical insemination (ICI), intrauterine insemination (IUI) or for in-vitro fertilization (IVF) or assisted reproduction technologies (ART).

Following the necessary quarantine period, which is usually 6 months, a sample will be thawed and used to artificially inseminate a woman or used for another assisted reproduction technologies (ART) treatment.

Services

Use

Subject to any regulations restricting who can obtain donor sperm, donor sperm is available to all women who, for whatever reason, want or need them. These regulations vary significantly between jurisdictions, and some countries do not have any regulations. When a woman finds that she is barred from receiving donor sperm within her jurisdiction, she may travel to another jurisdiction to obtain sperm. Regulations change from time to time. In general, donor sperm is readily available to a woman if her partner is infertile or where he has a genetic disorder, and the categories of women who may obtain donor sperms is expanding, with its availability to single women and to lesbian couples being now more widely available. Increasingly, donor sperm is used to achieve a pregnancy where a woman has no male partner, such as lesbian and bisexual women, and some sperm banks supply fertility centres which specialise in the treatment of such women.. Men may also store their own sperm at a sperm bank for future use particularly where they anticipate traveling to a war zone or having to undergo chemotherapy which might damage the testes.

Sperm from a sperm donor may also be used in surrogacy arrangements and for creating embryos for embryo donation. Donor sperm may be supplied by the sperm bank directly to the recipient to enable a woman to perform her own artificial insemination which can be carried out using a needleless syringe or a cervical cap conception device. The cervical cap conception device allows the donor semen to be held in place close to the cervix for between six and eight hours to allow fertilization to take place. Alternatively, donor sperm can be supplied by a sperm bank through a registered medical practitioner who will perform an appropriate method of insemination or IVF treatment using the donor sperm in order for the woman to become pregnant.

Choosing Donors

Information About Donor

In the United States, sperm banks maintain lists or catalogs of donors which provide basic information about the donor such as racial origin, skin color, height, weight, colour of eyes, and blood group. Some of these catalogs are available for browsing on the Internet, while others are made available to patients only when they apply to a sperm bank for treatment. Some sperm banks make additional information about each donor available for an additional fee, and others make additional basic information known to children produced from donors when those children reach the age of 18. Some clinics offer "exclusive donors" whose sperm is used to produce pregnancies for only one recipient woman. How accurate this is, or can be, is not known, and neither is it known whether the information produced by sperm banks, or by the donors themselves, is true. Many sperm banks will, however, carry out whatever checks they can to verify the information they request, such as checking the identity of the donor and contacting his own doctor to verify medical details.

In the United Kingdom, most donors are anonymous at the point of donation and recipi-ents can only non-identifying information about their donor (height, weight, ethnici-ty etc.). Donors need to provide identifying information to the clinic and clinics will usu-ally ask the donor's doctor to confirm any medical details they have been given. Donors are asked to provide a pen portrait of themselves which is held by the HFEA and can be obtained by the adult conceived from the donation at the age of 18, along with identify-ing information such as the donor's name and last known address. Known donation is permitted and it is not uncommon for family or friends to donate to a recipient couple.

Qualities that potential recipients typically prefer in donors include the donors being tall, college educated, and with a consistently high sperm count. A review came to the result that 68% of donors had given information to the clinical staff regarding physical characteristics and education but only 16% had provided additional information such as hereditary aptitudes and temperament or character.

Recipient's Selection of Donors

Sperm banks make information available about the sperm donors whose donations they hold to enable customers to select the donor whose sperm they wish to use. This information is often available by way of an on-line catalog. A sperm bank will also usually have facilities to help customers to make their choice and they will be able to advise on the suitability of donors for individual donors and their partners.

Where the recipient woman has a partner she may prefer to use sperm from a donor whose physical features are similar to those of her partner. In some cases, the choice of a donor with the correct blood group will be paramount, with particular considerations for the protection of recipients with negative blood groups. If a surrogate is to be used, such as where the customer is not intending to carry the child, considerations about her blood group etc. will also need to be taken into account.

Information made available by a sperm bank will usually include the race, height, weight, blood group, health and eye colour of the donor. Sometimes information about his age, family history and educational achievements will also be given. Some sperm banks make a 'personal profile' of a donor available and occasionally more information may be purchased about a donor, either in the form of a DVD or in written form. Catalogs usually state whether samples supplied in respect of a particular donor have already given rise to pregnancies, but this is not necessarily a guide to the fecundity of the sperm since a donor may not have been in the program long enough for any pregnancies to have been recorded.

If a woman intends to have more than one child, she may wish to have the additional child or children by the same donor. Sperm banks will usually advise whether sufficient stocks of sperm are available from a particular donor for subsequent pregnancies, and they normally have facilities available so that the woman may purchase and store additional vials from that donor on payment of an appropriate fee. These will be stored until required for subsequent pregnancies or they may be onsold if they become surplus to the woman's requirements.

The catalogue will also state whether samples of sperm are available for ICI, IUI, or IVF use.

Sex Selection

Some sperm banks enable recipients to choose the sex of their child, through methods of sperm sorting. Although the methods used do not guarantee 100% success, the chances of being able to select the gender of a child are held to be considerably increased.

One of the processes used is the 'swim up' method, whereby a sperm extender is added to the donor's freshly ejaculated sperm and the test-tube is left to settle. After about half-an-hour, the lighter sperm, containing the male chromosome pair (XY), will have swum to the top, leaving the heavier sperm, containing the female chromosome pair (XX), at the bottom, thus allowing selection and storage according to sex.

The alternative process is the Percoll Method which is similar to the 'swim up' method but involves additionally the centrifuging of the sperm in a similar way to the washing of samples produced for IUI inseminations, or for IVF purposes.

Sex selection is not permitted in a number of countries, including the UK.

Other Uses

A sperm bank may onsell sperm stocks to another entity instead of using it in fertility treatments.

There is a market for vials of processed sperm and for various reasons a sperm bank may sell-on stocks of vials which it holds (known as 'onselling'). Onselling enables a sperm bank to maximize the sale and disposal of sperm samples which it has processed. The reasons for onselling may be where part of, or even the main business of, a particular sperm bank is to process and store sperm rather than to use it in fertility treatments, or where a sperm bank is able to collect and store more sperm than it can use within nationally set limits. In the latter case a sperm bank may onsell sperm from a particular donor for use in another jurisdiction after the number of pregnancies achieved from that donor has reached its national maximum.

Sperm banks may supply other sperm banks or a fertility clinic with donor sperm to be used for achieving pregnancies.

Sperm banks may also supply sperm for research or educational purposes.

Regulation

In the United States, sperm banks are regulated as Human Cell and Tissue or Cell and Tissue Bank Product (HCT/Ps) establishments by the Food and Drug Administration (FDA) with new guidelines in effect May 25, 2005. Many states also have regulations in addition to those imposed by the FDA, including New York and California.

In the European Union a sperm bank must have a license according to the EU Tissue Directive which came into effect on April 7, 2006. In the United Kingdom, sperm banks are regulated by the Human Fertilisation and Embryology Authority.

In countries where sperm banks are allowed to operate, the sperm donor will not usually become the legal father of the children produced from the sperm he donates, but he will be the 'biological father' of such children. In cases of surrogacy involving embryo donation, a form of 'gestational surrogacy', the 'commissioning mother' or the 'commissioning parents' will not be biologically related to the child and may need to go through an adoption procedure.

As with other forms of third party reproduction, the use of donor sperm from a sperm bank gives rise to a number of moral, legal and ethical issues.

Semen Extender

Semen extender is a liquid diluent which is added to semen to preserve its fertilizing ability. It acts as a buffer to protect the sperm cells from their own toxic byproducts, and it protects the sperm cells from cold shock and osmotic shock during the chilling and shipping process (the sperm is chilled to reduce metabolism and allow it to live longer). The extender allows the semen to be shipped to the female, rather than requiring the male and female to be near to each other. Special freezing extender use also allows cryogenic preservation of sperm ("frozen semen"), which may be transported for use, or used on-site at a later date.

Semen extenders should not be confused with drugs or nutritional supplements designed to increase the volume of semen released during an ejaculation. The efficacy and utility of such products is dubious.

Function

The addition of extender to semen protects the sperm against possible damage by toxic seminal plasma, as well as providing nutrients and cooling buffers if the semen is to be cooled. In the case of freezing extenders, one or more penetrating cryoprotectants will be added. Typical cryoprotectants include glycerol, DMSO and dimethylformamide. Egg yolk, which has cryoprotective properties, is also a common component.

Ingredients

In the equine, Kenney extender (named after its developer, Dr. Robert M. Kenney of the University of Pennsylvania, New Bolton) has been used for many years, and contains a non-fat dried milk solid (NFDMS) and glucose. Dual-sugar extenders typically have similar ingredients, with an additional sugar, sucrose. Other extenders (e.g. INRA '96) may also contain milk components.

Antibiotics are almost universal in semen extenders, especially those being used for shipping or freezing semen. Ticarcillin (often used in combination with clavulanic acid under the designation timentin), amikacin sulfate, penicillin, and gentamicin are commonly used. The latter - gentamicin - has been noted to reduce sperm motility in the equine. In human semen extenders, antibiotics are required for regulatory reasons, so their use is almost universal in clinics, even though antibiotics can be detrimental to sperm. This is because in procedures such as IVF with frozen sperm the sperm do not need to swim up the reproductive tract on their own, and the detrimental effects of the antibiotics are not problematic. When private donors ship chilled semen outside of the formal regulatory environment, and fertilization is accomplished by allowing sperm to swim through the reproductive tract without the help of procedures such as IVF, then it is possible to achieve better results without antibiotics.

Brands

Porcine

- AndroPRO® Plus MOFA Global
- EnduraGuard® Plus MOFA Global
- ApX2® Plus MOFA Global
- AndroPRO® AM/PM MOFA Global
- Androhep Plus Minitube MOFA Global
- Androstar Plus Minitube
- MIII Minitube
- BTS Minitube MOFA Global
- MR-A Kubus
- Beltsville Liquid (BL-1)
- Acromax
- Beltsville Thawing Solution (BTS)
- Illinois Variable Temperature (IVT)
- Modena MOFA Global
- Kiev
- MULBERRY III
- Reading
- X-Cell
- Zorlesco
- ZORPVA
- MS Dilufert 9

- MS Dilufert 10
- MS Dilufert 3
- MS Dilufert 6
- Preserv Xtra

Bovine

- BoviPRO® CryoGuard MOFA Global
- Bioxcell
- CRYOBOS
- Andromed Minitube
- Andromed CSS Minitube
- Triladyl Minitube
- Biladyl Minitube
- Steridyl Minitube
- Biociphos

Equine

- EquiPRO® CoolGuard MOFA Global
- EquiPRO® CryoGuard MOFA Global
- Kenney (often marketed under a brand name - e.g. E-Z Mixin')
- VMDZ
- INRA '96
- Universal dual-sugar
- AndroMed-E Minitube
- Gent Extenders Minitube
- Equipro Minitube
- Spervital extenders (http://www.spervital.nl)
- EQUIDIL (EMBRYOLAB-UFSM)

Human

- Test Yolk Buffer (TYB) by Irvine
- Spermprep TYB

In Vitro Fertilisation

In vitro fertilisation (or fertilization; IVF) is a process by which an egg is fertilised by sperm outside the body: *in vitro* ("in glass"). The process involves monitoring and stimulating a woman's ovulatory process, removing an ovum or ova (egg or eggs) from the woman's ovaries and letting sperm fertilise them in a liquid in a laboratory. The fertilised egg (zygote) is cultured for 2–6 days in a growth medium and is then transferred to the same or another woman's uterus, with the intention of establishing a successful pregnancy.

IVF techniques can be used in different types of situations. It is a technique of assisted reproductive technology for treatment of infertility. IVF techniques are also employed in gestational surrogacy, in which case the fertilised egg is implanted into a surrogate's uterus, and the resulting child is genetically unrelated to the surrogate. In some situations, donated eggs or sperms may be used. Some countries ban or otherwise regulate the availability of IVF treatment, giving rise to fertility tourism. Restrictions on availability of IVF include costs and age to carry a healthy pregnancy to term. Due to the costs of the procedure, IVF is mostly attempted only after less expensive options have failed.

The first successful birth of a "test tube baby", Louise Brown, occurred in 1978. Louise Brown was born as a result of natural cycle IVF where no stimulation was made. Robert G. Edwards was awarded the Nobel Prize in Physiology or Medicine in 2010, the physiologist who co-developed the treatment together with Patrick Steptoe; Steptoe was not eligible for consideration as the Nobel Prize is not awarded posthumously. With egg donation and IVF, women who are past their reproductive years or menopause can still become pregnant. Adriana Iliescu held the record as the oldest woman to give birth using IVF and donated egg, when she gave birth in 2004 at the age of 66, a record passed in 2006. After the IVF treatment many couples are able to get pregnant without any fertility treatments. In 2012 it was estimated that five million children had been born worldwide using IVF and other assisted reproduction techniques.

Terminology

The term *in vitro*, from the Latin meaning *in glass*, is used, because early biological experiments involving cultivation of tissues outside the living organism from which they came, were carried out in glass containers such as *beakers, test tubes, or petri dishes.*

Today, the scientific term *in vitro* is used to refer to any biological procedure that is performed outside the organism in which it would normally have occurred, to distinguish it from an *in vivo* procedure, where the tissue remains inside the living organism within which it is normally found. A colloquial term for babies conceived as the result of IVF, "test tube babies", refers to the tube-shaped containers of glass or plastic resin, called *test tubes,* that are commonly used in chemistry labs and biology labs. However, *in vitro* fertilisation is usually performed in the shallower containers called Petri dishes. One IVF method, autologous endometrial coculture, is actually performed on organic material, but is still considered *in vitro*.

Medical Uses

IVF may be used to overcome female infertility where it is due to problems with the fallopian tubes, making fertilisation *in vivo* difficult. It can also assist in male infertility, in those cases where there is a defect in sperm quality; in such situations intracytoplasmic sperm injection (ICSI) may be used, where a sperm cell is injected directly into the egg cell. This is used when sperm has difficulty penetrating the egg, and in these cases the partner's or a donor's sperm may be used. ICSI is also used when sperm numbers are very low. When indicated, the use of ICSI has been found to increase the success rates of IVF.

According to the British NICE guidelines, IVF treatment is appropriate in cases of unexplained infertility for women that have not conceived after 2 years of regular unprotected sexual intercourse. This rule does not apply to all countries.

IVF is also considered suitable in cases where any of its expansions is of interest, that is, a procedure that is usually not necessary for the IVF procedure itself, but would be virtually impossible or technically difficult to perform without concomitantly performing methods of IVF. Such expansions include preimplantation genetic diagnosis (PGD) to rule out presence of genetic disorders, as well as egg donation or surrogacy where the woman providing the egg isn't the same who will carry the pregnancy to term. Further details in the Expansions-section below.

Success Rates

IVF success rates are the percentage of all IVF procedures which result in a favourable outcome. Depending on the type of calculation used, this outcome may represent the number of confirmed pregnancies, called the pregnancy rate, or the number of live births, called the live birth rate. The success rate depends on variable factors such as maternal age, cause of infertility, embryo status, reproductive history and lifestyle factors.

Maternal age: Younger candidates of IVF are more likely to get pregnant. Women older than 41 are more likely to get pregnant with a donor egg.

Reproductive history: Women who have been previously pregnant are in many cases more successful with IVF treatments then those who have never been pregnant.

Due to advances in reproductive technology, IVF success rates are substantially higher today than they were just a few years ago.

Live Birth Rate

The live birth rate is the percentage of all IVF cycles that lead to a live birth. This rate does not include miscarriage or stillbirth and multiple-order births such as twins and triplets are counted as one pregnancy. A 2012 summary compiled by the Society for Reproductive Medicine which reports the average IVF success rates in the United States per age group using non-donor eggs compiled the following data:

	<35	35-37	38-40	41-42	>42
Pregnancy rate	46.7	37.8	29.7	19.8	8.6
Live birth rate	40.7	31.3	22.2	11.8	3.9

In 2006, Canadian clinics reported a live birth rate of 27%. Birth rates in younger patients were slightly higher, with a success rate of 35.3% for those 21 and younger, the youngest group evaluated. Success rates for older patients were also lower and decrease with age, with 37-year-olds at 27.4% and no live births for those older than 48, the oldest group evaluated. Some clinics exceeded these rates, but it is impossible to determine if that is due to superior technique or patient selection, because it is possible to artificially increase success rates by refusing to accept the most difficult patients or by steering them into oocyte donation cycles (which are compiled separately). Further, pregnancy rates can be increased by the placement of several embryos at the risk of increasing the chance for multiples.

The live birth rates using donor eggs are also given by the SART and include all age groups using either fresh or thawed eggs.

	Fresh donor egg embryos	Thawed donor egg embryos
Live birth rate	55.1	33.8

Because not each IVF cycle that is started will lead to oocyte retrieval or embryo transfer, reports of live birth rates need to specify the denominator, namely IVF cycles started, IVF retrievals, or embryo transfers. The Society for Assisted Reproductive Technology (SART) summarised 2008-9 success rates for US clinics for fresh embryo cycles that did not involve donor eggs and gave live birth rates by the age of the prospective mother, with a peak at 41.3% per cycle started and 47.3% per embryo transfer for patients under 35 years of age.

IVF attempts in multiple cycles result in increased cumulative live birth rates. Depending on the demographic group, one study reported 45% to 53% for three attempts, and 51% to 71% to 80% for six attempts.

Pregnancy Rate

Pregnancy rate may be defined in various ways. In the United States, the pregnancy rate used by the Society for Assisted Reproductive Technology and the Centers for Disease Control (and appearing in the table in the Success Rates section above) are based on fetal heart motion observed in ultrasound examinations.

The 2009 summary compiled by the Society for Reproductive Medicine included the following data for the United States:

	<35	35-37	38-40	41-42
Pregnancy rate	47.6	38.9	30.1	20.5

In 2006, Canadian clinics reported an average pregnancy rate of 35%. A French study estimated that 66% of patients starting IVF treatment finally succeed in having a child (40% during the IVF treatment at the centre and 26% after IVF discontinuation). Achievement of having a child after IVF discontinuation was mainly due to adoption (46%) or spontaneous pregnancy (42%).

Predictors of Success

The main potential factors that influence pregnancy (and live birth) rates in IVF have been suggested to be maternal age, duration of infertility or subfertility, bFSH and number of oocytes, all reflecting ovarian function. Optimal woman's age is 23–39 years at time of treatment.

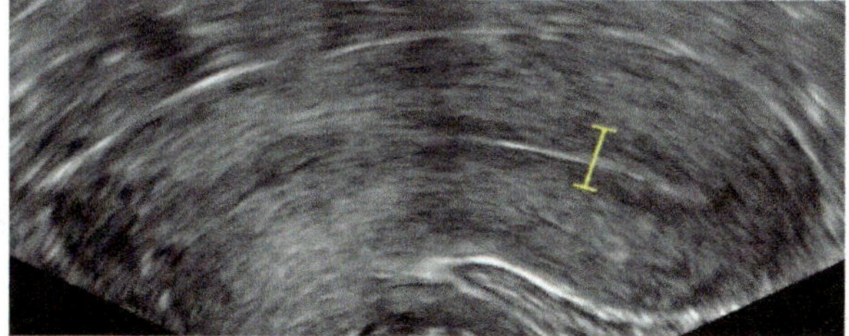

A triple-line endometrium is associated with better IVF outcomes.

Biomarkers that affect the pregnancy chances of IVF include:

- Antral follicle count, with higher count giving higher success rates.
- Anti-Müllerian hormone levels, with higher levels indicating higher chances of pregnancy, as well as of live birth after IVF, even after adjusting for age.
- Factors of semen quality for the sperm provider.
- Level of DNA fragmentation as measured e.g. by Comet assay, advanced maternal age and semen quality.

- Women with ovary-specific FMR1 genotypes including *het-norm/low* have significantly decreased pregnancy chances in IVF.
- Progesterone elevation (PE) on the day of induction of final maturation is associated with lower pregnancy rates in IVF cycles in women undergoing ovarian stimulation using GnRH analogues and gonadotrophins. At this time, compared to a progesterone level below 0.8 ng/ml, a level between 0.8 and 1.1 ng/ml confers an odds ratio of pregnancy of approximately 0.8, and a level between 1.2 and 3.0 ng/ml confers an odds ratio of pregnancy of between 0.6 and 0.7. On the other hand, progesterone elevation does not seem to confer a decreased chance of pregnancy in frozen–thawed cycles and cycles with egg donation.
- Characteristics of cells from the cumulus oophorus and the membrana granulosa, which are easily aspirated during oocyte retrieval. These cells are closely associated with the oocyte and share the same microenvironment, and the rate of expression of certain genes in such cells are associated with higher or lower pregnancy rate.
- An endometrial thickness (EMT) of less than 7 mm decreases the pregnancy rate by an odds ratio of approximately 0.4 compared to an EMT of over 7 mm. However, such low thickness rarely occurs, and any routine use of this parameter is regarded as not justified.

Other determinants of outcome of IVF include:

- Tobacco smoking reduces the chances of IVF producing a live birth by 34% and increases the risk of an IVF pregnancy miscarrying by 30%.
- A body mass index (BMI) over 27 causes a 33% decrease in likelihood to have a live birth after the first cycle of IVF, compared to those with a BMI between 20 and 27. Also, pregnant women who are obese have higher rates of miscarriage, gestational diabetes, hypertension, thromboembolism and problems during delivery, as well as leading to an increased risk of fetal congenital abnormality. Ideal body mass index is 19–30.
- Salpingectomy or laparoscopic tubal occlusion before IVF treatment increases chances for women with hydrosalpinges
- Success with previous pregnancy and/or live birth increases chances
- Low alcohol/caffeine intake increases success rate
- The number of embryos transferred in the treatment cycle.
- Embryo quality
- Some studies also suggest the autoimmune disease may also play a role in de-

creasing IVF success rates by interfering with proper implantation of the embryo after transfer.

Aspirin is sometimes prescribed to women for the purpose of increasing the chances of conception by IVF, but there is insufficient evidence to show that it actually works.

A 2013 review and metaanalysis of randomised controlled trials of acupuncture as an adjuvant therapy in IVF found no overall benefit, and concluded that an apparent benefit detected in a subset of published trials where the control group (those not using acupuncture) experienced a lower than average rate of pregnancy requires further study, due to the possibility of publication bias and other factors.

A Cochrane review came to the result that endometrial injury performed in the month prior to ovarian hyperstimulation appeared to increase both the live birth rate and clinical pregnancy rate in IVF compared with no endometrial injury. However, there was a lack of data reported on the rates of adverse outcomes such as miscarriage, multiple pregnancy, pain and/or bleeding.

For women, intake of antioxidants (such as N-acetyl-cysteine, melatonin, vitamin A, vitamin C, vitamin E, folic acid, myo-inositol, zinc or selenium) has not been associated with a significantly increased live birth rate or clinical pregnancy rate in IVF according to Cochrane reviews. On the other hand, oral antioxidants given to the men in couples with male factor or unexplained subfertility resulted in significantly higher live birth rate in IVF.

A Cochrane review in 2013 came to the result that there is no evidence identified regarding the effect of pre-conception lifestyle advice on the chance of a live birth outcome.

Complications

Multiple Births

The major complication of IVF is the risk of multiple births. This is directly related to the practice of transferring multiple embryos at embryo transfer. Multiple births are related to increased risk of pregnancy loss, obstetrical complications, prematurity, and neonatal morbidity with the potential for long term damage. Strict limits on the number of embryos that may be transferred have been enacted in some countries (e.g. Britain, Belgium) to reduce the risk of high-order multiples (triplets or more), but are not universally followed or accepted. Spontaneous splitting of embryos in the womb after transfer can occur, but this is rare and would lead to identical twins. A double blind, randomised study followed IVF pregnancies that resulted in 73 infants (33 boys and 40 girls) and reported that 8.7% of singleton infants and 54.2% of twins had a birth weight of < 2,500 grams (5.5 lb).

Recent evidence also suggest that singleton offspring after IVF is at higher risk for lower birth weight for unknown reasons.

Spread of Infectious Disease

By sperm washing, the risk that a chronic disease in the male providing the sperm would infect the female or offspring can be brought to negligible levels.

In males with hepatitis B, The Practice Committee of the American Society for Reproductive Medicine advises that sperm washing is not necessary in IVF to prevent transmission, unless the female partner has not been effectively vaccinated. In females with hepatitis B, the risk of vertical transmission during IVF is no different from the risk in spontaneous conception. However, there is not enough evidence to say that ICSI procedures are safe in females with hepatitis B in regard to vertical transmission to the offspring.

Regarding potential spread of HIV/AIDS, Japan's government prohibited the use of *in vitro* fertilisation procedures for couples in which both partners are infected with HIV. Despite the fact that the ethics committees previously allowed the Ogikubo, Tokyo Hospital, located in Tokyo, to use *in vitro* fertilisation for couples with HIV, the Ministry of Health, Labour and Welfare of Japan decided to block the practice. Hideji Hanabusa, the vice president of the Ogikubo Hospital, states that together with his colleagues, he managed to develop a method through which scientists are able to remove HIV from sperm.

Other Risks to the Egg Provider/Retriever

A risk of ovarian stimulation is the development of ovarian hyperstimulation syndrome, particularly if hCG is used for inducing final oocyte maturation. This results in swollen, painful ovaries. It occurs in 30% of patients. Mild cases can be treated with over the counter medications and cases can be resolved in the absence of pregnancy. In moderate cases, ovaries swell and fluid accumulated in the abdominal cavities and may have symptoms of heartburn, gas, nausea or loss of appetite. In severe cases patients have sudden excess abdominal pain, nausea, vomiting and will result in hospitalisation.

During egg retrieval, there's a small chance of bleeding, infection, and damage to surrounding structures like bowel and bladder (transvaginal ultrasound aspiration) as well as difficulty in breathing, chest infection, allergic reactions to medication, or nerve damage (laproscopy).

Ectopic pregnancy may also occur if a fertilised egg develops outside the uterus, usually in the fallopian tubes and requires immediate destruction of the fetus.

IVF does not seem to be associated with an elevated risk of cervical cancer, nor with ovarian cancer or endometrial cancer when neutralising the confounder of infertility itself. Nor does it seem to impart any increased risk for breast cancer.

Regardless of pregnancy result, IVF treatment is usually stressful for patients. Neuroticism and the use of escapist coping strategies are associated with a higher degree of distress, while the presence social support has a relieving effect. A negative pregnancy test after IVF is associated with an increased risk for depression in women, but not with any increased risk of developing anxiety disorders. Pregnancy test results do not seem to be a risk factor for depression or anxiety among men.

Birth Defects

A review in 2013 came to the result that infants resulting from IVF (with or without ICSI) have a relative risk of birth defects of 1.32 (95% confidence interval 1.24–1.42) compared to naturally conceived infants. In 2008, an analysis of the data of the National Birth Defects Study in the US found that certain birth defects were significantly more common in infants conceived through IVF, notably septal heart defects, cleft lip with or without cleft palate, esophageal atresia, and anorectal atresia; the mechanism of causality is unclear. However, in a population-wide cohort study of 308,974 births (with 6163 using assisted reproductive technology and following children from birth to age five) researchers found: "The increased risk of birth defects associated with IVF was no longer significant after adjustment for parental factors." Parental factors included known independent risks for birth defects such as maternal age, smoking status, etc. Multivariate correction did not remove the significance of the association of birth defects and ICSI (corrected odds ratio 1.57), although the authors speculate that underlying male infertility factors (which would be associated with the use of ICSI) may contribute to this observation and were not able to correct for these confounders. The authors also found that a history of infertility elevated risk itself in the absence of any treatment (odds ratio 1.29), consistent with a Danish national registry study and "...implicates patient factors in this increased risk." The authors of the Danish national registry study speculate: "...our results suggest that the reported increased prevalence of congenital malformations seen in singletons born after assisted reproductive technology is partly due to the underlying infertility or its determinants."

Risk in singleton pregnancies resulting from IVF (with or without ICSI)

Condition	Relative risk	95% confidence interval
Beckwith–Wiedemann syndrome	3-4	
congenital anomalies	1.67	1.33–2.09
ante-partum haemorrhage	2.49	2.30–2.69
hypertensive disorders of pregnancy	1.49	1.39–1.59
preterm rupture of membranes	1.16	1.07–1.26
Caesarean section	1.56	1.51–1.60
gestational diabetes	1.48	1.33–1.66
induction of labour	1.18	1.10–1.28
small for gestational age	1.39	1.27–1.53

preterm birth	1.54	1.47–1.62
low birthweight	1.65	1.56–1.75
perinatal mortality	1.87	1.48–2.37

Other Risks to the Offspring

If the underlying infertility is related to abnormalities in spermatogenesis, it is plausible, but too early to examine that male offspring are at higher risk for sperm abnormalities.

IVF does not seem to confer any risks regarding cognitive development, school performance, social functioning and behaviour. Also, IVF infants are known to be as securely attached to their parents as those who were naturally conceived, and IVF adolescents are as well-adjusted as those who have been naturally conceived.

Limited long-term follow-up data suggest that IVF may be associated with an increased incidence of hypertension, impaired fasting glucose, increase in total body fat composition, advancement of bone age, subclinical thyroid disorder, early adulthood clinical depression and binge drinking in the offspring. It is not known, however, whether these potential associations are caused by the IVF procedure in itself, by adverse obstetric outcomes associated with IVF, by the genetic origin of the children or by yet unknown IVF-associated causes. Increases in embryo manipulation during IVF result in more deviant fetal growth curves, but birth weight does not seem to be a reliable marker of fetal stress.

IVF, including ICSI, is associated with an increased risk of imprinting disorders (including Prader-Willi syndrome and Angelman syndrome), with an odds ratio of 3.7 (95% confidence interval 1.4 to 9.7).

An IVF-associated incidence of cerebral palsy and neurodevelopmental delay are believed to be related to the confounders of prematurity and low birthweight. Similarly, an IVF-associated incidence of autism and attention-deficit disorder are believed to be related to confounders of maternal and obstetric factors.

Overall, IVF does not cause an increased risk of childhood cancer. Studies have shown a decrease in the risk of certain cancers and an increased risks of certain others including retinoblastoma hepatoblastoma and rhabdomyosarcoma.

Method

Theoretically, *in vitro* fertilisation could be performed by collecting the contents from a woman's fallopian tubes or uterus after natural ovulation, mixing it with sperm, and reinserting the fertilised ova into the uterus. However, without additional techniques, the chances of pregnancy would be extremely small. The additional techniques that are routinely used in IVF include ovarian hyperstimulation to generate multiple eggs or ultrasound-guided transvaginal oocyte retrieval directly from the ovaries; after which the ova and sperm are prepared, as well as culture and selection of resultant embryos before embryo transfer into a uterus.

Ovarian Hyperstimulation

Ovarian hyperstimulation is the stimulation to induce development of multiple follicles of the ovaries. It should start with response prediction by e.g. age, antral follicle count and level of anti-Müllerian hormone. The resulting prediction of e.g. poor or hyper-response to ovarian hyperstimulation determines the protocol and dosage for ovarian hyperstimulation.

Ovarian hyperstimulation also includes suppression of spontaneous ovulation, for which two main methods are available: Using a (usually longer) GnRH agonist protocol or a (usually shorter) GnRH antagonist protocol. In a standard long GnRH agonist protocol the day when hyperstimulation treatment is started and the expected day of later oocyte retrieval can be chosen to conform to personal choice, while in a GnRH antagonist protocol it must be adapted to the spontaneous onset of the previous menstruation. On the other hand, the GnRH antagonist protocol has a lower risk of ovarian hyperstimulation syndrome (OHSS), which is a life-threatening complication.

For the ovarian hyperstimulation in itself, injectable gonadotropins (usually FSH analogues) are generally used under close monitoring. Such monitoring frequently checks the estradiol level and, by means of gynecologic ultrasonography, follicular growth. Typically approximately 10 days of injections will be necessary.

Natural IVF

There are several methods termed *natural cycle IVF*:

- IVF using no drugs for ovarian hyperstimulation, while drugs for ovulation suppression may still be used.
- IVF using ovarian hyperstimulation, including gonadotropins, but with a GnRH antagonist protocol so that the cycle initiates from natural mechanisms.
- Frozen embryo transfer; IVF using ovarian hyperstimulation, followed by embryo cryopreservation, followed by embryo transfer in a later, natural, cycle.

IVF using no drugs for ovarian hyperstimulation was the method for the conception of Louise Brown. This method can be successfully used when women want to avoid taking ovarian stimulating drugs with its associated side-effects. HFEA has estimated the live birth rate to be approximately 1.3% per IVF cycle using no hyperstimulation drugs for women aged between 40–42.

Mild IVF is a method where a small dose of ovarian stimulating drugs are used for a short duration during a woman's natural cycle aimed at producing 2–7 eggs and creating healthy embryos. This method appears to be an advance in the field to reduce complications and side-effects for women and it is aimed at quality, and not quantity of eggs and embryos. One study comparing a mild treatment (mild ovarian stimulation with GnRH antagonist co-treatment combined with single embryo transfer) to a standard treatment (stimulation with a GnRH agonist long-protocol and transfer of two embryos) came to the result that the proportions of cumulative pregnancies that resulted in term live birth after 1 year were 43.4% with mild treatment and 44.7% with standard treatment. Mild IVF can be cheaper than conventional IVF and with a significantly reduced risk of multiple gestation and OHSS.

Final Maturation Induction

When the ovarian follicles have reached a certain degree of development, induction of final oocyte maturation is performed, generally by an injection of human chorionic gonadotropin (hCG). Commonly, this is known as the "trigger shot." hCG acts as an analogue of luteinising hormone, and ovulation would occur between 38 and 40 hours after a single HCG injection, but the egg retrieval is performed at a time usually between 34 and 36 hours after hCG injection, that is, just prior to when the follicles would rupture. This avails for scheduling the egg retrieval procedure at a time where the eggs are fully mature. HCG injection confers a risk of ovarian hyperstimulation syndrome. Using a GnRH agonist instead of hCG eliminates the risk of ovarian hyperstimulation syndrome, but with a delivery rate of approximately 6% less than with hCG.

Egg Retrieval

The eggs are retrieved from the patient using a transvaginal technique called transvaginal oocyte retrieval, involving an ultrasound-guided needle piercing the vaginal wall to reach the ovaries. Through this needle follicles can be aspirated, and the follicular fluid is passed to an embryologist to identify ova. It is common to remove between ten and thirty eggs. The retrieval procedure usually takes between 20 and 40 minutes, depending on the number of mature follicles, and is usually done under conscious sedation or general anaesthesia.

Egg and Sperm Preparation

In the laboratory, the identified eggs are stripped of surrounding cells and prepared for fertilisation. An oocyte selection may be performed prior to fertilisation to select eggs with optimal chances of successful pregnancy. In the meantime, semen is prepared for fertilisation by removing inactive cells and seminal fluid in a process called sperm washing. If semen is being provided by a sperm donor, it will usually have been prepared for treatment before being frozen and quarantined, and it will be thawed ready for use.

Co-incubation

The sperm and the egg are incubated together at a ratio of about 75,000:1 in a culture media in order for the actual fertilisation to take place. A review in 2013 came to the result that a duration of this co-incubation of about 1 to 4 hours results in significantly higher pregnancy rates than 16 to 24 hours. In most cases, the egg will be fertilised during co-incubation and will show two pronuclei. In certain situations, such as low sperm count or motility, a single sperm may be injected directly into the egg using intracytoplasmic sperm injection (ICSI). The fertilised egg is passed to a special growth medium and left for about 48 hours until the egg consists of six to eight cells.

In gamete intrafallopian transfer, eggs are removed from the woman and placed in one of the fallopian tubes, along with the man's sperm. This allows fertilisation to take place inside the woman's body. Therefore, this variation is actually an *in vivo* fertilisation, not an *in vitro* fertilisation.

Embryo Culture

The main durations of embryo culture are until cleavage stage (day two to four after co-incubation) or the blastocyst stage (day five or six after co-incubation). Embryo culture until the blastocyst stage confers a significant increase in live birth rate per embryo transfer, but also confers a decreased number of embryos available for transfer and embryo cryopreservation, so the cumulative clinical pregnancy rates are increased with cleavage stage transfer. Transfer day two instead of day three after fertilisation has no differences in live birth rate. There are significantly higher odds of preterm birth (odds ratio 1.3) and congenital anomalies (odds ratio 1.3) among births having from embryos cultured until the blastocyst stage compared with cleavage stage.

Embryo Selection

Laboratories have developed grading methods to judge oocyte and embryo quality. In order to optimise pregnancy rates, there is significant evidence that a morphological scoring system is the best strategy for the selection of embryos. Since 2009 where the first time-lapse microscopy system for IVF was approved for clinical use, morphokinetic scoring systems has shown to improve to pregnancy rates further. However, when all different types of time-lapse embryo imaging devices, with or without morphokinetic scoring systems, are compared against conventional embryo assessment for IVF, there is insufficient evidence of a difference in live-birth, pregnancy, stillbirth or miscarriage to choose between them.

Embryo Transfer

Embryos are graded by the embryologist based on the amount of cells, evenness of growth and degree of fragmentation. The number to be transferred depends on the number available, the age of the woman and other health and diagnostic factors. In countries such as Canada, the UK, Australia and New Zealand, a maximum of two embryos are transferred except in unusual circumstances. In the UK and according to HFEA regulations, a woman over 40 may have up to three embryos transferred, whereas in the USA, younger women may have many embryos transferred based on individual fertility diagnosis. Most clinics and country regulatory bodies seek to minimise the risk of pregnancies carrying multiples, as it is not uncommon for more implantations to take than desired. The embryos judged to be the "best" are transferred to the patient's uterus through a thin, plastic catheter, which goes through her vagina and cervix. Several embryos may be passed into the uterus to improve chances of implantation and pregnancy.

Adjunctive Medication

Luteal support is the administration of medication, generally progesterone, progestins or GnRH agonists, to increase the success rate of implantation and early embryogenesis, thereby complementing and/or supporting the function of the corpus luteum. The live birth rate is significantly higher with progesterone for luteal support in IVF cycles with or without intracytoplasmic sperm injection (ICSI). Co-treatment with GnRH agonists further improves outcomes, by a live birth rate RD of +16% (95% confidence interval +10 to +22%).

On the other hand, growth hormone or aspirin as adjunctive medication in IVF have no evidence of overall benefit.

Expansions

There are various expansions or additional techniques that can be applied in IVF, which are usually not necessary for the IVF procedure itself, but would be virtually impossible or technically difficult to perform without concomitantly performing methods of IVF.

Preimplantation Genetic Screening or Diagnosis

Preimplantation genetic screening (PGS) or preimplantation genetic diagnosis (PGD) has been suggested to be able to be used in IVF to select an embryo that appears to have the greatest chances for successful pregnancy. However, a systematic review and meta-analysis of existing randomised controlled trials came to the result that there is no evidence of a beneficial effect of PGS with cleavage-stage biopsy as measured by live birth rate. On the contrary, for women of advanced maternal age, PGS with cleavage-stage biopsy significantly lowers the live birth rate. Technical drawbacks, such as the invasiveness of the biopsy, and non-representative samples because of mosaicism are the major underlying factors for inefficacy of PGS.

Still, as an expansion of IVF, patients who can benefit from PGS/PGD include:

- Couples who have a family history of inherited disease

- Couples who want to use gender selection to prevent a gender-linked disease
- Couples who already have a child with an incurable disease and need compatible cells from a second healthy child to cure the first, resulting in a "saviour sibling" that matches the sick child in HLA type.

PGS screens for numeral chromosomal abnormalities while PGD diagnosis the specific molecular defect of the inherited disease. In both PGS and PGD, individual cells from a pre-embryo, or preferably trophectoderm cells biopsied from a blastocyst, are analysed during the IVF process. Before the transfer of a pre-embryo back to a woman's uterus, one or two cells are removed from the pre-embryos (8-cell stage), or preferably from a blastocyst. These cells are then evaluated for normality. Typically within one to two days, following completion of the evaluation, only the normal pre-embryos are transferred back to the woman's uterus. Alternatively, a blastocyst can be cryopreserved via vitrification and transferred at a later date to the uterus. In addition, PGS can significantly reduce the risk of multiple pregnancies because fewer embryos, ideally just one, are needed for implantation.

Cryopreservation

Cryopreservation can be performed as oocyte cryopreservation before fertilisation, or as embryo cryopreservation after fertilisation.

The Rand Consulting Group has estimated there to be 400,000 frozen embryos in the United States. The advantage is that patients who fail to conceive may become pregnant using such embryos without having to go through a full IVF cycle. Or, if pregnancy occurred, they could return later for another pregnancy. Spare oocytes or embryos resulting from fertility treatments may be used for oocyte donation or embryo donation to another woman or couple, and embryos may be created, frozen and stored specifically for transfer and donation by using donor eggs and sperm. Also, oocyte cryopreservation can be used for women who are likely to lose their ovarian reserve due to undergoing chemotherapy.

The outcome from using cryopreserved embryos has uniformly been positive with no increase in birth defects or development abnormalities.

Other Expansions

- Intracytoplasmic sperm injection (*ICSI*) is where a single sperm is injected directly into an egg. Its main usage as an expansion of IVF is to overcome male infertility problems, although it may also be used where eggs cannot easily be penetrated by sperm, and occasionally in conjunction with sperm donation. It can be used in teratozoospermia, since once the egg is fertilised abnormal sperm morphology does not appear to influence blastocyst development or blastocyst morphology.

- Additional methods of embryo profiling. For example, methods are emerging in making comprehensive analyses of up to entire genomes, transcriptomes, proteomes and metabolomes which may be used to score embryos by comparing the patterns with ones that have previously been found among embryos in successful versus unsuccessful pregnancies.

- Assisted zona hatching (AZH) can be performed shortly before the embryo is transferred to the uterus. A small opening is made in the outer layer surrounding the egg in order to help the embryo hatch out and aid in the implantation process of the growing embryo.

- In egg donation and embryo donation, the resultant embryo after fertilisation is inserted in another woman than the one providing the eggs. These are resources for women with no eggs due to surgery, chemotherapy, or genetic causes; or with poor egg quality, previously unsuccessful IVF cycles or advanced maternal age. In the egg donor process, eggs are retrieved from a donor's ovaries, fertilised in the laboratory with the sperm from the recipient's partner, and the resulting healthy embryos are returned to the recipient's uterus.

- In oocyte selection, the oocytes with optimal chances of live birth can be chosen. It can also be used as a means of preimplantation genetic screening.

- Embryo splitting can be used for twinning to increase the number of available embryos.

- Cytoplasmic transfer is where the cytoplasm from a donor egg is injected into an egg with compromised mitochondria. The resulting egg is then fertilized with sperm and implanted in a womb, usually that of the woman who provided the recipient egg and nuclear DNA. Cytoplasmic transfer was created to aid women who experience infertility due to deficient or damaged mitochondria, contained within an egg's cytoplasm.

Leftover Embryos or Eggs

There may be leftover embryos or eggs from IVF procedures if the woman for whom they were originally created has successfully carried one or more pregnancies to term. With the woman's or couple's permission, these may be donated to help other women or couples as a means of third party reproduction.

In embryo donation, these extra embryos are given to other couples or women for transfer with the goal of producing a successful pregnancy. The resulting child is considered the child of the woman who carries it and gives birth, and not the child of the donor, the same as occurs with egg donation or sperm donation.

Typically, genetic parents donate the eggs to a fertility clinic orwhere they are preserved by oocyte cryopreservation or embryo cryopreservation until a carrier is found for them. Typically the process of matching the embryo(s) with the prospective parents is conducted by the agency itself, at which time the clinic transfers ownership of the embryos to the prospective parents.

In the United States, women seeking to be an embryo recipient undergo infectious disease screening required by the U.S. Food and Drug Administration (FDA), and reproductive tests to determine the best placement location and cycle timing before the actual Embryo Transfer occurs. The amount of screening the embryo has already undergone is largely dependent on the genetic parents' own IVF clinic and process. The embryo recipient may elect to have her own embryologist conduct further testing.

Alternatives to donating unused embryos are destroying them (or having them implanted at a time where pregnancy is very unlikely), keeping them frozen indefinitely, or donating them for use in research (which results in their unviability). Individual moral views on disposing leftover embryos may depend on personal views on the beginning of human personhood and definition and/or value of potential future persons and on the value that is given to fundamental research questions. Some people believe donation of leftover embryos for research is a good alternative to discarding the embryos when patients receive proper, honest and clear information about the research project, the procedures and the scientific values.

History

In 1977, Steptoe and Edwards successfully carried out a pioneering conception which resulted in the birth of the world's first baby to be conceived by IVF, Louise Brown on 25 July 1978, in Oldham General Hospital, Greater Manchester, UK.

The second successful birth of a test tube baby occurred in India just 67 days after Louise Brown was born. The girl, named Durga conceived *in vitro* using the methods of Subhash Mukhopadhyay, a physician and researcher from Kolkata.

Ethics

Mix-ups

In some cases, laboratory mix-ups (misidentified gametes, transfer of wrong embryos) have occurred, leading to legal action against the IVF provider and complex paternity suits. An example is the case of a woman in California who received the embryo of another couple and was notified of this mistake after the birth of her son. This has led to many authorities and individual clinics implementing procedures to minimise the risk of such mix-ups. The HFEA, for example, requires clinics to use a double witnessing system, the identity of specimens is checked by two people at each point at which specimens are transferred. Alternatively, technological solutions are gaining favour, to reduce the manpower cost of manual double witnessing, and to further reduce risks with uniquely numbered RFID tags which can be identified by readers connected to a computer. The computer tracks specimens throughout the process and alerts the embryologist if non-matching specimens are identified. Although the use of RFID tracking has expanded in the USA, it is still not widely adopted. However, In other cases there has been not mix-up of embryos or gametes, but the intentional use of embryos of another couple or gamete donor, without informed consent of parents, both: receptors or donors. Some of these cases are taking a legal and judicial course.

Preimplantation Genetic Diagnosis or Screening

Another concern is that people will screen in or out for particular traits, using preimplantation genetic diagnosis (PGD) or preimplantation genetic screening. For example, a deaf British couple, Tom and Paula Lichy, have petitioned to create a deaf baby using IVF. Some medical ethicists have been very critical of this approach. Jacob M. Appel wrote that "intentionally culling out blind or deaf embryos might prevent considerable future suffering, while a policy that allowed deaf or blind parents to select *for* such traits intentionally would be far more troublesome."

This concept of decisively altering genes has coined the concept of the Designer Baby. Currently, PGD can alter some physical and health attributes, and projections for the future power of PGD in its ability to create the ideal human has raised many ethical issues. Projections for societal repercussions include changing the realm athletics, creating human weapons, and exchanging autonomy over one's life course for predesignation. Also, with a limited view of the future, it is difficult to alter a human's genetic makeup without knowing full repercussions. For example, through gene therapy, a lab was able to make rats lose weight, but the long-term effects of the gene manipulation lead to worry of toxin production and too much weight loss. To prevent some of these issues from arising, scientists work towards stabilising the entire process to make it safer before applying a higher degree of gene modification to the human embryos in IVF.

Autonomy and Tissue Ownership

For those who believe that human life begins at the moment of conception, this belief also suggests that human rights are given at that time. If human rights are given in this embryonic stage, then a surplus of ethical issues arise from manipulating the embryo in the realm of tissue ownership. In the long run, if implanted into a female and born, the embryo becomes an adult and has to now live with the genetic modifications chosen for them through the process of IVF. Unfortunately, in this base, cellular state, consent for gene manipulation is impossible. This leads to decision making by the parents. Rightful parental ownership over the embryo is only in the short-run and means that they control the embryos biological future. Consent over tissue ownership has been an issue for decades and can have legal repercussions. In the case of Henrietta Lacks, researchers lacked patient consent to use her tissues in genetic research, and this led to many legal issues on the family's right to profit from the use of her cells. Decisiveness over autonomy is necessary in the case of IVF to avoid long run issues and give people their full rights of humanity.

Profit Desire of the Industry

Many people do not oppose the IVF practice itself (i.e. the creating of a pregnancy through "artificial" ways) but are highly critical of the current state of the present day industry. Such individuals argue that the industry has now become a multibillion-dollar industry, which is widely unregulated and prone to serious abuses in the desire of practitioners to obtain profit. For instance, in 2008, a California physician transferred 12 embryos to a woman who gave birth to octuplets. This has made international news, and had led to accusations that many doctors are willing to seriously endanger the health and even life of women in order to gain money. Robert Winston, professor of fertility studies at Imperial College London, had called the industry "corrupt" and "greedy" saying that "One of the major problems facing us in healthcare is that IVF has become a massive commercial industry," and that "What has happened, of course, is that money is corrupting this whole technology", and accused authorities of failing to protect couples from exploitation "The regulatory authority has done a consistently bad job. It's not prevented the exploitation of women, it's not put out very good information to couples, it's not limited the number of unscientific treatments people have access to". The IVF industry can thus be seen as an example of what social scientists are describing as an increasing trend towards a market-driven construction of health, medicine and the human body.

As the science progresses, the industry is further driven by money in that researchers and innovators enter into the fight over patents and intellectual property rights. The Copyright Clause in the US Constitution protects innovator's rights to their respective work in attempts to promote scientific progress. Essentially, this lawful protection gives incentive to the innovators by providing them a temporary monopoly over their respective work. In the IVF industry, already incredibly expensive for patients, patents risk even higher prices for the patients to receive the procedure as they have to also cover the costs of protected works. For example, company 23andMe has patented a process used to calculate probability of gene inheritance. While this innovation could help many, the company retains sole right to administer it and thus does not have economic competition. Lack of economic competition leads to higher prices of products.

The industry has been accused of making unscientific claims, and distorting facts relating to infertility, in particular through widely exaggerated claims about how common infertility is in society, in an attempt to get as many couples as possible and as soon as possible to try treatments (rather than trying to conceive naturally for a longer time). This risks removing infertility from its social context and reducing the experience to a simple biological malfunction, which not only *can* be treated through bio-medical procedures, but *should* be treated by them. Indeed, there are serious concerns about the overuse of treatments, for instance Dr Sami David, a fertility specialist and one of the pioneers of the early days of the IVF treatments, has expressed disappointment over the current state of the industry, and said many procedures are unnecessary; he said: "It's being the first choice of treatment rather than the last choice. When it was first opening up in late 1970s, early 80s, it was meant to be the last resort. Now it's a first resort. I think that's an injustice to women. I also think it can harm women in the long run." IVF thus raises ethical issues concerning the abuse of bio-medical facts to 'sell' corrective procedures and treatments for conditions that deviate from a constructed ideal of the 'healthy' or 'normal' body i.e., fertile females and males with reproductive systems capable of co-producing offspring.

Pregnancy Past Menopause

Although menopause is a natural barrier to further conception, IVF has allowed women to be pregnant in their fifties and sixties. Women whose uteruses have been appropriately prepared receive embryos that originated from an egg of an egg donor. Therefore, although these women do not have a genetic link with the child, they have an emotional link through pregnancy and childbirth. In many cases the genetic father of the child is the woman's partner. Even after menopause the uterus is fully capable of carrying out a pregnancy.

Allowing women to get pregnant past the natural time can factor into issues of overpopulation. Through the PGD, children born through IVF would credibly have higher life expectancy rates due to eliminated diseases. So increasing the amount of women who are able to bear children increases the population growth rate, while PGD in IVF decreases the death rate, resulting in an increasing population.

Same-sex Couples, Single and Unmarried Parents

A 2009 statement from the ASRM found no persuasive evidence that children are harmed or disadvantaged solely by being raised by single parents, unmarried parents, or homosexual parents. It did not support restricting access to assisted reproductive technologies on the basis of a prospective parent's marital status or sexual orientation.

Ethical concerns include reproductive rights, the welfare of offspring, nondiscrimination against unmarried individuals, homosexual, and professional autonomy.

A recent controversy in California focused on the question of whether physicians opposed to same-sex relationships should be required to perform IVF for a lesbian couple. Guadalupe T. Benitez, a lesbian medical assistant from San Diego, sued doctors Christine Brody and Douglas Fenton of the North Coast Women's Care Medical Group after Brody told her that she had "religious-based objections to treating her and homosexuals in general to help them conceive children by artificial insemination," and Fenton refused to authorise a refill of her prescription for the fertility drug Clomid on the same grounds. The California Medical Association had initially sided with Brody and Fenton, but the case, North Coast Women's Care Medical Group v. Superior Court, was decided unanimously by the California State Supreme Court in favour of Benitez on 19 August 2008.

Nadya Suleman came to international attention after having twelve embryos implanted, eight of which survived, resulting in eight newborns being added to her existing six-child family. The Medical Board of California sought to have fertility doctor Michael Kamrava, who treated Suleman, stripped of his licence. State officials allege that performing Suleman's procedure is evidence of unreasonable judgment, substandard care, and a lack of concern for the eight children she would conceive and the six she was already struggling to raise. On 1 June 2011 the Medical Board issued a ruling that Kamrava's medical licence be revoked effective 1 July 2011.

Anonymous Donors

Some children conceived by IVF using anonymous donors report being troubled over not knowing about their donor parent as well any genetic relatives they may have and their family history.

Alana Stewart, who was conceived using donor sperm, began an online forum for donor children called AnonymousUS in 2010. The forum welcomes the viewpoints of anyone involved in the IVF process. Olivia Pratten, a donor-conceived Canadian, sued the province of British Columbia for access to records on her donor father's identity in 2008. "I'm not a treatment, I'm a person, and those records belong to me," Pratten said. In May 2012, a court ruled in Pratten's favour, agreeing that the laws at the time discriminated against donor children and making anonymous sperm and egg donation in British Columbia illegal.

In the U.K., Sweden, Norway, Germany, Italy, New Zealand, and some Australian states, donors are not paid and cannot be anonymous.

In 2000, a website called Donor Sibling Registry was created to help biological children with a common donor connect with each other.

In 2012, a documentary called *Anonymous Father's Day* was released that focuses on donor-conceived children.

Unwanted Embryos

During the selection and transfer phases many embryos may be discarded in favour of others. This selection may be based on criteria such as genetic disorders or the sex. One of the earliest cases of special gene selection through IVF was the case of the Collins family in the 1990s, who selected the sex of their child. The ethic issues remain unresolved as no consensus exists in science, religion, and philosophy on when a human embryo should be recognised as a person. For those who believe that this is at the moment of conception, IVF becomes a moral question when multiple eggs are fertilised, begin development, and only a few are chosen for implantation.

If IVF were to involve the fertilisation of only a single egg, or at least only an amount that will be implanted, then this would not be an issue. However, this has the chance of increasing costs dramatically as only a few eggs can be attempted at a time. As a result, the couple must decide what to do with these extra embryos. Depending on their view of the embryo's humanity or the chance the couple will want to try to have another child, the couple has multiple options for dealing with these extra embryos. Couples can choose to keep them frozen, donate them to other infertile couples, thaw them, or donate them to medical research. Keeping them frozen costs money, donating them does not ensure they will survive, thawing them renders them immediately unviable, and medical research results in their termination. In the realm of medical research, the couple is not necessarily told what the embryos will be used for, and as a result, some can be used in stem cell research, a field perceived to have ethical issues.

Religious Response

The Roman Catholic Church opposes all kinds of assisted reproductive technology and artificial contraception, asserting they separate the procreative goal of marital sex from the goal of uniting married couples. The Roman Catholic Church permits the use of a small number of reproductive technologies and contraceptive methods like natural family planning, which involves charting ovulation times. The church allows other forms of reproductive technologies that allow conception to take place from normative sexual intercourse, such as a fertility lubricant. Pope Benedict XVI had publicly re-emphasised the Catholic Church's opposition to *in vitro* fertilisation, claiming it replaces love between a husband and wife. The Catechism of the Catholic Church claims that Natural law teaches that reproduction has an "inseparable connection" to sexual union of married couples. In addition, the church opposes IVF because it might cause disposal of embryos; in Catholicism, an embryo is viewed as an individual with a soul that must be treated as a person. The Catholic Church maintains that it is not objectively evil to be infertile, and advocates adoption as an option for such couples who still wish to have children.

Hindus welcomed the IVF as gift for those who can't bear child and termed doctors related to IVF doing punya as there are several characters who were claimed to be born without intercourse, mainly Karna and five Pandavas.

Regarding the response to IVF of Islam, the conclusions of Gad El-Hak Ali Gad El-Hak's ART fatwa include that:

- IVF of an egg from the wife with the sperm of her husband and the transfer of the fertilised egg back to the uterus of the wife is allowed, provided that the procedure is indicated for a medical reason and is carried out by an expert physician.
- Since marriage is a contract between the wife and husband during the span of their marriage, no third party should intrude into the marital functions of sex and procreation. This means that a third party donor is not acceptable, whether he or she is providing sperm, eggs, embryos, or a uterus. The use of a third party is tantamount to *zina*, or adultery.

Within the Orthodox Jewish community the concept is debated as there is little precedent in traditional Jewish legal textual sources. Regarding laws of sexuality, religious challenges include masturbation (which may be regarded as "seed wasting"), laws related to sexual activity and menstruation (niddah) and the specific laws regarding intercourse. An additional major issue is that of establishing paternity and lineage. For a baby conceived naturally, the father's identity is determined by a legal presumption (chazakah) of legitimacy: *rov bi'ot achar ha'baal* - a woman's sexual relations are assumed to be with her husband. Regarding an IVF child, this assumption does not exist and as such Rabbi Eliezer Waldenberg (among others) requires an outside supervisor to positively identify the father. Reform Judaism has generally approved *in vitro* fertilisation.

Society and Culture

Many people of sub-Saharan Africa choose to foster their children to infertile women. IVF enables these infertile women to have their own children, which imposes new ideals to a culture in which fostering children is seen as both natural and culturally important. Many infertile women are able to earn more respect in their society by taking care of the children of other mothers, and this may be lost if they choose to use IVF instead. As IVF is seen as unnatural, it may even hinder their societal position as opposed to making them equal with fertile women. It is also economically advantageous for infertile women to raise foster children as it gives these children greater ability to access resources that are important for their development and also aids the development of their society at large. If IVF becomes more popular without the birth rate decreasing, there could be more large family homes with fewer options to send their newborn children. This could result in an increase of orphaned children and/or a decrease in resources for the children of large families. This would ultimately stifle the children's and the community's growth.

Emotional Involvement

Studies have indicated that IVF mothers show greater emotional involvement with their child, and they enjoy motherhood more than mothers by natural conception. Similarly, studies have indicated that IVF fathers express more warmth and emotional involvement than fathers by adoption and natural conception and enjoy fatherhood more. Some IVF parents become overly involved with their children.

Men and IVF

Research has shown that men largely view themselves as 'passive' contributors since they have 'less physical involvement' in IVF treatment. Despite this, many men feel distressed after seeing the toll of hormonal injections and ongoing physical intervention on their partner. Fertility was found to be a significant factor in a man's perception of his masculinity, driving many to keep the treatment a secret. In cases where the men did share that he and his partner were undergoing IVF, they reported to have been teased, mainly by other men, although some viewed this as an affirmation of support and friendship. For others, this led to feeling socially isolated. In comparison with women, men showed less deterioration in mental health in the years following a failed treatment. However many men did feel guilt, disappointment and inadequacy, stating that they were simply trying to provide an 'emotional rock' for their partners.

Availability and Utilisation

High costs keep IVF out of reach for many developing countries, but research by the Genk Institute for Fertility Technology, in Belgium, claim to have found a much lower cost methodology (about 90% reduction) with similar efficacy, which may be suitable for some fertility treatment. Moreover, the laws of many countries permit IVF for only single women, lesbian couples, and persons participating in surrogacy arrangements. Using PGD gives members of these select demographic groups disproportionate access to a means of creating a child possessing characteristics that they consider "ideal," raising issues of equal opportunity for both the parents'/parent's and the child's generation. Many fertile couples now demand equal access to embryonic screening so that their child can be just as healthy as one created through IVF. Mass use of PGD, especially as a means of population control or in the presence of legal measures related to population or demographic control, can lead to intentional or unintentional demographic effects such as the skewed live-birth sex ratios seen in communist China following implementation of its one-child policy.

USA

In the USA, overall availability of IVF in 2005 was 2.5 IVF physicians per 100,000 population, and utilisation was 236 IVF cycles per 100,000. Utilisation highly increases with availability and IVF insurance coverage, and to a significant extent also with percentage of single persons and median income. In the USA 126 procedures are performed per million people per year. In the USA an average cycle, from egg retrieval to embryo implantation, costs $12,400, and insurance companies that do cover treatment, even partially, usually cap the number of cycles they pay for.

The cost of IVF rather reflects the costliness of the underlying healthcare system than the regulatory or funding environment, and ranges, on average for a standard IVF cycle and in 2006 United States dollars, between $12,500 in the United States to $4,000 in Japan. In Ireland, IVF costs around €4,000, with fertility drugs, if required, costing up to €3,000. The cost per live birth is highest in the United States ($41,000) and United Kingdom ($40,000) and lowest in Scandinavia and Japan (both around $24,500).

Many fertility clinics in the United States limit the upper age at which women are eligible for IVF to 50 or 55 years. These cut-offs make it difficult for women older than fifty-five to utilise the procedure.

Australia

In Australia, the average age of women undergoing ART treatment is 35.5 years among those using their own eggs (one in four being 40 or older) and 40.5 years among those using donated eggs.

Cameroon

Ernestine Gwet Bell supervised the first Cameroonian child born through IVF in 1998.

Israel

Israel has the highest rate of IVF in the world, with 1657 procedures performed per million people per year. The second highest rate is in Iceland, with 899 procedures per million people per year. Israel provides unlimited free in vitro procedures for its citizens for up to two children per woman under 45 years of age. In other countries the coverage of such procedures is limited if it exists at all. The Israeli Health Ministry says it spends roughly $3450 per procedure.

United Kingdom

Availability of IVF in England is determined by Clinical commissioning groups. The National Institute for Health and Care Excellence recommends up to 3 cycles of treatment for women under 40 and one cycle for some women aged between 40 and 42, but financial pressure has eroded compliance with this recommendation. CCGs in Essex, Bedfordshire and Somerset have reduced funding to one cycle, or none and it is expected that reductions will become more widespread. Funding may be available in "exceptional circumstances" – for example if a male partner has a transmittable infection or one partner is affected by cancer treatment. According to the campaign group Fertility Fairness at the end of 2014 every CCG in England was funding at least one cycle of IVF". Prices paid by the NHS in England varied between under £3,000 to more than £6,000 in 2014/5.

Legal Status

Government agencies in China passed bans on the use of IVF in 2003 by unmarried women or by couples with certain infectious diseases.

Sunni Muslim nations generally allow IVF between married couples when conducted with their own respective sperm and eggs, but not with donor eggs from other couples. But Iran, which is Shi'a Muslim, has a more complex scheme. Iran bans sperm donation but allows donation of both fertilised and unfertilised eggs. Fertilised eggs are donated from married couples to other married couples, while unfertilised eggs are donated in the context of mut'ah or temporary marriage to the father.

Costa Rica has a complete ban on IVF technology, it having been ruled unconstitutional by the nation's Supreme Court because it "violated life." Costa Rica is the only country in the western hemisphere that forbids IVF. A law project sent reluctantly by the government of President Laura Chinchilla was rejected by parliament. President Chinchilla has not publicly stated her position on the question of IVF. However, given the massive influence of the Catholic Church in her government any change in the status quo seems very unlikely. In spite of Costa Rican government and strong religious opposition, the IVF ban has been struck down by the Inter-American Court of Human Rights in a decision of 20 December 2012. The court said that a long-standing Costa Rican guarantee of protection for every human embryo violated the reproductive freedom of infertile couples because it prohibited them from using IVF, which often involves the disposal of embryos not implanted in a patient's uterus. On 10 September 2015, President Luis Guillermo Solís signed a decree legalising in-vitro fertilisation. The decree was added to the country's official gazette on 11 September. Opponents of the practice have since filed a lawsuit before the country's Constitutional Court.

All major restrictions on single but infertile women using IVF were lifted in Australia in 2002 after a final appeal to the Australian High Court was rejected on procedural grounds in the Leesa Meldrum case. A Victorian federal court had ruled in 2000 that the existing ban on all single women and lesbians using IVF constituted sex discrimination. Victoria's government announced changes to its IVF law in 2007 eliminating remaining restrictions on fertile single women and lesbians, leaving South Australia as the only state maintaining them.

Federal regulations in the United States include screening requirements and restrictions on donations, but generally do not affect sexually intimate partners. However, doctors may be required to *provide* treatments due to nondiscrimination laws, as for example in California. The US state of Tennessee proposed a bill in 2009 that would have defined donor IVF as adoption. During the same session another bill proposed barring adoption from any unmarried and cohabitating couple, and activist groups stated that passing the first bill would effectively stop unmarried people from using IVF. Neither of these bills passed.

Embryo Transfer

Embryo transfer refers to a step in the process of assisted reproduction in which embryos are placed into the uterus of a female with the intent to establish a pregnancy. This technique (which is often used in connection with in vitro fertilization (IVF)), may be used in humans or in animals, in which situations the goals may vary.

Embryo transfer	
Intervention	
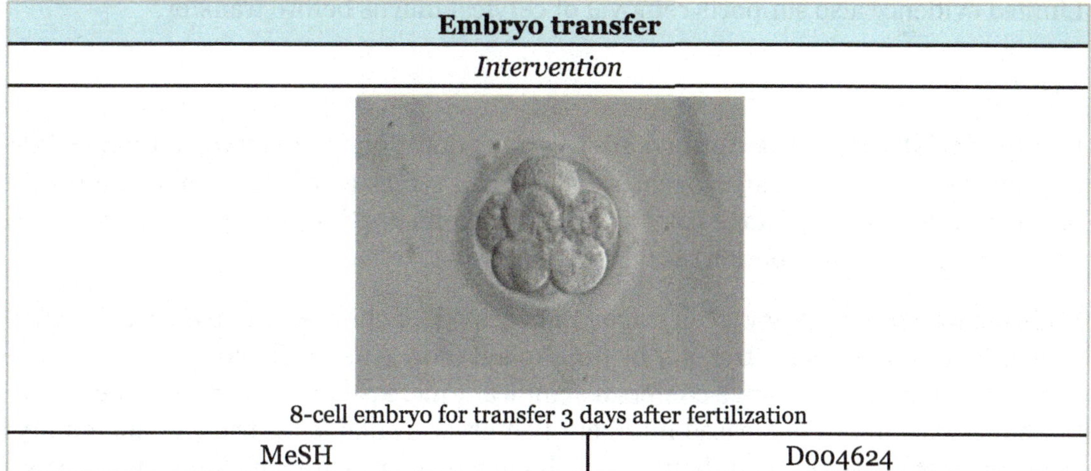 8-cell embryo for transfer 3 days after fertilization	
MeSH	D004624

Fresh Versus Frozen

Embryos can be either "fresh" from fertilized egg cells of the same menstrual cycle, or "frozen", that is they have been generated in a preceding cycle and undergone embryo cryopreservation, and are thawed just prior to the transfer, which is then termed "frozen embryo transfer" (FET). The outcome from using cryopreserved embryos has uniformly been positive with no increase in birth defects or development abnormalities, also between fresh versus frozen eggs used for intracytoplasmic sperm injection (ICSI). In fact, pregnancy rates are increased following FET, and perinatal outcomes are less affected, compared to embryo transfer in the same cycle as ovarian hyperstimulation was performed. The endometrium is believed to not be optimally prepared for implantation following ovarian hyperstimulation, and therefore frozen embryo transfer avails for a separate cycle to focus on optimizing the chances of successful implantation. Children born from vitrified blastocysts have significantly higher birthweight than those born from non-frozen blastocysts. When transferring a frozen-thawed oocyte, the chance of pregnancy is essentially the same whether it is transferred in a natural cycle or one with ovulation induction.

Uterine Preparation

In the human, the uterine lining (endometrium) needs to be appropriately prepared so that the embryo(s) can implant. In a natural cycle the embryo transfer takes place in the luteal phase at a time where the lining is appropriately undeveloped in relation to the status of the present Luteinizing Hormone. In a stimulated or a cycle where a "frozen" embryo is transferred, the recipient woman could be given first estrogen preparations (about 2 weeks), then a combination of oestrogen and progesterone so that the lining becomes receptive for the embryo. The time of receptivity is the implantation window. A scientific review in 2013 came to the conclusion that it is not possible to identify one method of endometrium preparation in frozen embryo transfer as being more effective than another.

Limited evidence also supports removal of cervical mucus before transfer.

Timing

Embryo transfer can be performed after various durations of embryo culture, conferring different stages in embryogenesis. The main stages at which embryo transfer is performed are cleavage stage (day 2 to 4 after co-incubation) or the blastocyst stage (day 5 or 6 after co-incubation).

Embryos who reach the day 3 cell stage can be tested for chromosomal or specific genetic defects prior to possible transfer by preimplantation genetic diagnosis (PGD). Transferring at the blastocyst stage confers a significant increase in live birth rate per transfer, but also confers a decreased number of embryos available for transfer and embryo cryopreservation, so the cumulative clinical pregnancy rates are increased with cleavage stage transfer. Transfer day 2 instead of day 3 after fertilization has no differences in live birth rate.

Monozygotic twinning is not increased after blastocyst transfer compared with cleavage-stage embryo transfer.

There is a significantly higher odds of preterm birth (odds ratio 1.3) and congenital anomalies (odds ratio 1.3) among births having reached the blastocyst stage compared with cleavage stage.

Procedure

The embryo transfer procedure starts by placing a speculum in the vagina to visualize the cervix, which is cleansed with saline solution or culture media. A soft transfer catheter is loaded with the embryos and handed to the clinician after confirmation of the patient's identity. The catheter is inserted through the cervical canal and advanced into the uterine cavity.

There is good and consistent evidence of benefit in *ultrasound guidance*, that is, making an abdominal ultrasound to ensure correct placement, which is 1–2 cm from the uterine fundus. There is evidence of a significant increase in clinical pregnancy using ultrasound guidance compared with only "clinical touch". Anesthesia is generally not required. Single embryo transfers in particular require accuracy and precision in placement within the uterine cavity. The optimal target for embryo placement, known as the maximal implantation potential (MIP) point, is identified using 3D/4D ultrasound. However, there is limited evidence that supports deposition of embryos in the midportion of the uterus.

After insertion of the catheter, the contents are expelled and the embryos are deposited. Limited evidence supports making trial transfers before performing the procedure with embryos. After expulsion, the duration that the catheter remains inside the uterus has no effect on pregnancy rates. Limited evidence suggests avoiding negative pressure from the catheter after expulsion. After withdrawal, the catheter is handed to the embryologist, who inspects it for retained embryos.

In the process of zygote intrafallopian transfer (ZIFT), eggs are removed from the woman, fertilised, and then placed in the woman's fallopian tubes rather than the uterus.

Procedure Simulation

In 2015, the American Society for Reproductive Medicine developed a medical simulation of the Embryo Transfer procedure with Swiss company VirtaMed, designed for the education and training of clinicians. The virtual reality simulator, which includes real-time simulation of ultrasound guidance, was launched at the annual conference of the American Society for Reproductive Medicine.

Embryo Number

A major issue is how many embryos should be transferred. Placement of multiple embryos carries the risk of multiple pregnancy. In the past, physicians have often placed too many embryos in the hope to establish a pregnancy. However, the rise in multiple pregnancies has led to a reassessment of this approach. Professional societies and in many countries, the legislature, have issued guidelines or laws to curtail a practice of placing too many embryos in an attempt to reduce multiple pregnancies. The appropriate number of embryos to be transferred depends on the age of the woman, whether it is the first, second or third full IVF cycle attempt and whether there are top-quality embryos available. According to a guideline from The National Institute for Health and Care Excellence (NICE) in 2013, the number of embryos transferred in a cycle should be chosen as in following table:

Age	Attempt No.	Embryos transferred
<37 years	1st	1
	2nd	1 if top-quality
	3rd	No more than 2
37–39 years	1st & 2nd	1 if top-quality
		2 if no top-quality
	3rd	No more than 2
40–42 years		2

e-SET

The technique of selecting only one embryo to transfer to the woman is called elective-Single Embryo Transfer (e-SET) or, when embryos are at the blastocyst stage, it can also be called *elective single blastocyst transfer (eSBT)*. It significantly lowers the risk of multiple pregnancies, compared with e.g. Double Embryo Transfer (DET) or *double blastocyst transfer* (2BT), with a twinning rate of approximately 3.5% in sET compared with approximately 38% in DET, or 2% in eSBT compared with approximately 25% in 2BT. At the same time, pregnancy rates is not significantly less with eSBT than with 2BT. That is, the cumulative live birth rate associated with single fresh embryo transfer followed by a single frozen and thawed embryo transfer is comparable with that after one cycle of double fresh embryo transfer. Furthermore, SET has better outcomes in terms of mean gestational age at delivery, mode of delivery, birthweight, and risk of neonatal intensive care unit necessity than DET. e-SET of embryos at the cleavage stage reduces the likelihood of live birth by 38% and multiple birth by 94%. Evidence from randomized, controlled trials suggests that increasing the number of e-SET attempts (fresh and/or frozen) results in a cumulative live birth rate similar to that of DET.

The usage of single embryo transfer is highest in Sweden (69.4%), but as low as 2.8% in the USA. Access to public funding for ART, availability of good cryopreservation facilities, effective education about the risks of multiple pregnancy, and legislation appear to be the most important factors for regional usage of single embryo transfer. Also, personal choice plays a significant role as many subfertile couples have a strong preference for twins.

Adjunctive Procedures

There is limited evidence to support the use of mechanical closure of the cervical canal following embryo transfer.

There is insufficient evidence to support a certain amount of time for women to remain recumbent following embryo transfer.

There is no evidence of benefit in terms of live birth rate of using hyaluronic acid as adherence medium for the embryo. Neither is there any evidence of benefit of having a full bladder, removal of cervical mucus, or flushing of the endometrial or endocervical cavity at the time of embryo transfer. Adjunctive antibiotics in the form of amoxicillin plus clavulanic acid does not increase the clinical pregnancy rate compared with no antibiotics.

For frozen-thawed embryo transfer or transfer of embryo from egg donation, no previous ovarian hyperstimulation is required for the recipient before transfer, which can be performed in spontaneous ovulatory cycles. Still, various protocols exist for frozen-thawed embryo transfers as well, such as protocols with ovarian hyperstimulation, protocols in which the endometrium is artificially prepared by estrogen and/or progesterone. A Cochrane review in 2010 of randomized studies came to the result that there generally is insufficient evidence to support the use of one intervention in preference to another, but with some evidence that in cycles where the endometrium is artificially prepared by estrogen or progesterone, it is beneficial to administer an additional drug that suppresses hormone production by the ovaries such as continuous administration of a gonadotropin releasing hormone agonist (GnRHa). For egg donation, there is evidence of a lower pregnancy rate and a higher cycle cancellation rate when the progesterone supplementation in the recipient is commenced *prior* to oocyte retrieval from the donor, as compared to commenced day of oocyte retrieval or the day after.

Seminal fluid contains several proteins that interact with epithelial cells of the cervix and uterus, inducing active gestational immune tolerance. There are significantly improved outcomes when women are exposed to seminal plasma around the time of embryo transfer, with statistical significance for clinical pregnancy, but not for ongoing pregnancy or live birth rates with the limited data available.

Follow-up

Patients usually start progesterone medication after egg (also called oocyte) retrieval. While daily intramuscular injections of progesterone-in-oil (PIO) have been the standard route of administration, PIO injections are not FDA-approved for use in pregnancy. A recent meta-analysis showed that the intravaginal route with an appropriate dose and dosing frequency is equivalent to daily intramuscular injections. In addition, a recent case-matched study comparing vaginal progesterone with PIO injections showed that live birth rates were nearly identical with both methods. A duration of progesterone administration of 11 days results in almost the same birth rates as longer durations.

Patients are also given estrogen medication in some cases after the embryo transfer. Pregnancy testing is done typically two weeks after egg retrieval.

Third-party Reproduction

It is not necessary that the embryo transfer be performed on the female who provided the eggs. Thus another female whose uterus is appropriately prepared can receive the embryo and become pregnant. Embryo transfer may be used where a woman who has eggs but no uterus and wants to have a biological baby; she would require the help of a gestational carrier or surrogate to carry the pregnancy. Also, a woman who has no eggs but a uterus may utilize egg donor IVF, in which case another woman would provide eggs for fertilization and the resulting embryos are placed into the uterus of the patient. Fertilization may be performed using the woman's partner's sperm or by using donor sperm. 'Spare' embryos which are created for another couple undergoing IVF treatment but which are then surplus to that couple's needs may also be transferred (called embryo donation). Embryos may be specifically created by using eggs and sperm from donors and these can then be transferred into the uterus of another woman. A surrogate may carry a baby produced by embryo transfer for another couple, even though neither she nor the 'commissioning' couple is biologically related to the child. Third party reproduction is controversial and regulated in many countries. Persons entering gestational surrogacy arrangements must make sense of an entirely new type of relationship that does not fit any of the traditional scripts we use to categorize relations as kinship, friendship, romantic partnership or market relations. Surrogates have the experience of carrying a baby that they conceptualize as not of their own kin, while intended mothers have the experience of waiting through nine months of pregnancy and transitioning to motherhood from outside of the pregnant body. This can lead to new conceptualizations of body and self.

History

The first transfer of an embryo from one human to another resulting in pregnancy was reported in July 1983 and subsequently led to the announcement of the first human birth 3 February 1984. This procedure was performed at the Harbor UCLA Medical Center under the direction of Dr. John Buster and the University of California at Los Angeles School of Medicine.

In the procedure, an embryo that was just beginning to develop was transferred from one woman in whom it had been conceived by artificial insemination to another woman who gave birth to the infant 38 weeks later. The sperm used in the artificial insemination came from the husband of the woman who bore the baby.

This scientific breakthrough established standards and became an agent of change for women suffering from the afflictions of infertility and for women who did not want to pass on genetic disorders to their children. Donor embryo transfer has given women a mechanism to become pregnant and give birth to a child that will contain their husband's genetic makeup. Although donor embryo transfer as practiced today has evolved from the original non-surgical method, it now accounts for approximately 5% of in vitro fertilization recorded births.

Prior to this, thousands of women who were infertile, had adoption as the only path to parenthood. This set the stage to allow open and candid discussion of embryo donation and transfer. This breakthrough has given way to the donation of human embryos as a common practice similar to other donations such as blood and major organ donations. At the time of this announcement the event was captured by major news carriers and fueled healthy debate and discussion on this practice which impacted the future of reproductive medicine by creating a platform for further advancements in woman's health.

This work established the technical foundation and legal-ethical framework surrounding the clinical use of human oocyte and embryo donation, a mainstream clinical practice, which has evolved over the past 25 years. Building upon this groundbreaking research and since the initial birth announcement in 1984, well over 47,000 live births resulting from donor embryo transfer have been and continue to be recorded by the Centers for Disease Control(CDC) in the United States to infertile women, who otherwise would not have had children by any other existing method.

Effectiveness

A Cochrane systematic review updated in 2012 showed that blastocyst stage transfer is more effective than cleavage (day 2 or 3) stage transfer in assisted reproductive technologies. It showed a small improvement in live birth rate per couple for blastocyst transfers. This would mean that for a typical rate of 31% in clinics that use early cleavage stage cycles, the rate would increase to 32% to 42% live births if clinics used blastocyst transfer.

Embryo Transfer in Animals

Embryo transferred Charolais calves with their Angus and Hereford recipient mothers.

Embryo transfer techniques allow top quality female livestock to have a greater influence on the genetic advancement of a herd or flock in much the same way that artificial insemination has allowed greater use of superior sires. ET also allows the continued use of animals such as competition mares to continue training and showing, while producing foals. The general epidemiological aspects of embryo transfer indicates that the transfer of embryos provides the opportunity to introduce genetic material into populations of livestock while greatly reducing the risk for transmission of infectious diseases. Recent developments in the sexing of embryos before transfer and implanting has great potential in the dairy and other livestock industries.

Embryo transfer is also used in laboratory mice. For example, embryos of genetically modified strains that are difficult to breed or expensive to maintain may be stored frozen, and only thawed and implanted into a pseudopregnant dam when needed.

Frozen Embryo Transfer in Animals

The development of various methods of cryopreservation of bovine embryos improved embryo transfer technique considerably efficient technology, no longer depending on the immediate readiness of suitable recipients. Pregnancy rates are just slightly less than those achieved with fresh embryos. Recently, the use of cryoprotectants such as ethylene glycol has permitted the direct transfer of bovine embryos. The world's first live crossbred bovine calf produced under tropical conditions by Direct Transfer (DT) of embryo frozen in ethylene glycol freeze media was born on 23 June 1996.Dr.Binoy Sebastian Vettical of Kerala Livestock Development Board Ltd has produced the embryo stored frozen in Ethylene Glycol freeze media by slow programmable freezing (SPF) technique and transferred directly to recipient cattle immediately after thawing the frozen straw in water for the birth of this calf. In a study, in vivo produced crossbred bovine embryos stored frozen in ethylene glycol freeze media were transferred directly to recipients under tropical conditions and achieved a pregnancy rate of 50 percent. In a survey of the North American embryo transfer industry, embryo transfer success rates from direct transfer of embryos were as good as to those achieved with glycerol. Moreover, in 2011, more than 95% of frozen-thawed embryos were transferred by Direct Transfer.

Cryosurgery

Cryosurgery (cryotherapy) is the use of extreme cold in surgery to destroy abnormal or diseased tissue. The term comes from the Greek words cryo ("icy cold") and surgery (*cheirourgiki*) meaning "hand work" or "handiwork". Cryosurgery has been historically used to treat a number of diseases and disorders, especially a variety of benign and malignant skin conditions.

Uses

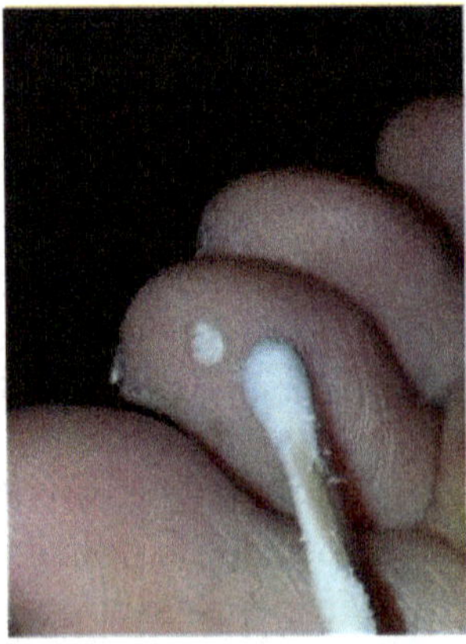

Cryotherapy to a plantar wart using cotton bud application

Warts, moles, skin tags, solar keratoses, Morton's neuroma and small skin cancers are candidates for cryosurgical treatment. Several internal disorders are also treated with cryosurgery, including liver cancer, prostate cancer, lung cancer, oral cancers, cervical disorders and, more commonly in the past, hemorrhoids. Soft tissue conditions such as plantar fasciitis (jogger's heel) and fibroma (benign excrescence of connective tissue) can be treated with cryosurgery. Generally, all tumors that can be reached by the cryoprobes used during an operation are treatable. Although found to be effective, this method of treatment is only appropriate for use against localized disease, and solid tumors larger than 1 cm. Tiny, diffuse metastases that often coincide with cancers are usually not affected by cryotherapy.

Cryosurgery works by taking advantage of the destructive force of freezing temperatures on cells. When their temperature sinks beyond a certain level ice crystals begin forming inside the cells and, because of their lower density, eventually tear apart those cells. Further harm to malignant growth will result once the blood vessels supplying the affected tissue begin to freeze.

Method

Liquid Nitrogen

Cryogun used to spray liquid nitrogen

A common method of freezing lesions is using liquid nitrogen as the cooling solution. This −196 °C (−321 °F) cold liquid may be sprayed on the diseased tissue, circulated through a tube called a cryoprobe, or simply dabbed on with a cotton or foam swab.

Carbon Dioxide

Carbon dioxide is also available as a spray and is used to treat a variety of benign spots. Less frequently, doctors use carbon dioxide "snow" formed into a cylinder or mixed with acetone to form a slush that is applied directly to the treated tissue.

Argon

Recent advances in technology have allowed for the use of argon gas to drive ice formation using a principle known as the Joule-Thomson effect. This gives physicians excellent control of the ice, and minimizing complications using ultra-thin 17 gauge cryoneedles.

Dimethyl Ether – propane

A mixture of dimethyl ether and propane is used in some preparations such as Dr. Scholl's Freeze Away. The mixture is stored in an aerosol spray type container at room temperature and drops to −41 °C (−42 °F) when dispensed. The mixture is often dispensed into a straw with a cotton-tipped swab.

Products

Cryosurgical Systems

A number of medical supply companies have developed cryogen delivery systems for cryosurgery. Most are based on the use of liquid nitrogen, although some employ the use of proprietary mixtures of gases that combine to form the cryogen. Some commonly used cryosurgical products are:

- Brymill
- Cry-Ac
- Cryoalfa
- CryoClear
- CryoPen
- CryoPro, Cortex Technology
- CryoProbe
- Cryosurgery, Inc. Verruca-Freeze
- Histofreezer
- MedGyn Cryotherapy System
- Miltex Cryosolutions
- Premier CryOmega
- Premier NitroSpray
- Myoscience Iovera

Cryosurgery In Cancer Treatment

Cryosurgery is also used to treat internal and external tumors as well as tumors in the bone. to cure internal tumors, a hollow instrument called a cryoprobe is used, which is placed in contact with the tumor. Liquid nitrogen or Argon gas is passed through that cryoprobe. Ultrasound or MRI is used to guide the cryoprobe and monitor the freezing of the cells. this helps in limiting damage to adjacent healthy tissues. A ball of ice crystals forms around the probe which results in freezing of nearby cells. when it is required to deliver gas to various parts of the tumor, more than one probe is used. After cryosurgery, the frozen tissue is either naturally absorbed by the body in case of internal tumors, or it dissolves and forms a scab for external tumors.

Results

Cryosurgery is a minimally invasive procedure, and is often preferred to more traditional kinds of surgery because of its minimal pain, scarring, and cost; however, as with any medical treatment, there are risks involved, primarily that of damage to nearby healthy tissue. Damage to nerve tissue is of particular concern.

Patients undergoing cryosurgery usually experience redness and minor-to-moderate localized pain, which most of the time can be alleviated sufficiently by oral administration of mild analgesics such as Ibuprofen, codeine, tramadol or acetaminophen (paracetamol). Blisters may form as a result of cryosurgery, but these usually scab over and peel away within a few days.

Cryoablation

Cryoablation is a process that uses extreme cold (cryo) to destroy or damage tissue (ablation).

Cryoablation is used in a variety of clinical applications using hollow needles (cryoprobes) through which cooled, thermally conductive, fluids are circulated. Cryoprobes are inserted into or placed adjacent to tissue which is determined to be diseased in such a way that ablation will provide correction yielding benefit to the patient. When the probes are in place, the cryogenic freezing unit removes heat ("cools") from the tip of the probe and by extension from the surrounding tissues.

Ablation occurs in tissue that has been frozen by at least three mechanisms:

1. formation of ice crystals within cells thereby disrupting membranes, and interrupting cellular metabolism among other processes;
2. coagulation of blood thereby interrupting bloodflow to the tissue in turn causing ischemia and cell death; and
3. induction of apoptosis, the so-called programmed cell death cascade.

The most common application of cryoablation is to ablate solid tumors found in the lung, liver, breast, kidney and prostate. The use in prostate and renal cryoablation are the most common. Although sometimes applied through laparoscopic or open surgical approaches, most often cryoablation is performed percutaneously (through the skin and into the target tissue containing the tumor) by a medical specialist, such as an interventional radiologist.

Prostate

Prostate cryoablation is moderately effective but, as with any prostate removal process, also can result in impotence. Prostate cryoablation is used in three patient categories:

1. as primary therapy in patients for whom sexual function is less important or who are poor candidates for radical retropubic prostatectomy (RRP, surgical removal of the prostate);
2. as salvage therapy in patients who have failed brachytherapy (the use of implanted radioactive "seeds" placed within the prostate) or external beam radiation therapy (EBRT); and
3. focal therapy for smaller, discrete tumors in younger patients.

Bone cancer

Cryoablation has been explored as a alternative to radiofrequency ablation in the treatment of moderate to severe pain in people with metastatic bone disease. The area of tissue destruction created by this technique can be monitored more effectively by CT than RFA, a potential advantage when treating tumors adjacent to critical structures.

Renal

Cryoablation has similar outcomes to radiofrequency ablation when treating renal cell carcinoma.

Breast Cancer

Cryoablation for breast cancer is typically only possible for small tumors. Often surgery is used following cryoablation. As of 2014 more research is required before it can replace lumpectomy.

Cardiac

Another type of cryoablation is used to restore normal electrical conduction by freezing tissue or heart pathways that interfere with the normal distribution of the heart's electrical impulses. Cryoablation is used in two types of intervention for the treatment of arrhythmias: (1) catheter-based procedures and (2) surgical operations.

A catheter is a very thin tube that is inserted into a vein in the patient's leg and threaded to the heart where it delivers energy to treat the patient's arrhythmia. In surgical procedures, a flexible probe is used directly on an exposed heart to apply the energy that interrupts the arrhythmia. By cooling the tip of a cryoablation catheter (cardiology) or probe (heart surgery) to sub-zero temperatures, the cells in the heart responsible for conducting the arrhythmia are altered so that they no longer conduct electrical impulses.

Fibroadenoma

Cryoablation is also currently being used to treat fibroadenomas of the breast. Fibroadenomas are benign breast tumors that are found in approximately 10% of women (primarily ages 15–30).

In this procedure which has been approved by the U.S. Food and Drug Administration (FDA), an ultrasound-guided probe is inserted into the fibroadenoma and extremely cold temperatures are then used to destroy the abnormal cells. Over time the cells are reabsorbed into the body. The procedure can be performed in a doctor's office setting with local anesthesia and leaves very little scarring compared to open surgical procedures.

Catheter-based Procedures

Different catheter-based ablation techniques may be used and they generally fall into two categories: (1) cold-based procedures where tissue cooling is used to treat the arrhythmia, and (2) heat-based procedures where high temperature is used to alter the abnormal conductive tissue in the heart.

Cryoablation

Cold temperatures are used in cryoablation to chill or freeze cells that conduct abnormal heart rhythms. The catheter removes heat from the tissue to cool it to temperatures as low as -75 °C. This causes localized scarring, which cuts undesired conduction paths.

This is a much newer treatment for supraventricular tachycardia (SVT) involving the atrioventicular (AV) node directly. SVT involving the AV node is often a contraindication for using radiofrequency ablation because of the risk of injuring the AV node, forcing patients to receive a permanent pacemaker. With cryoablation, areas of tissue can be mapped by limited, reversible, freezing (e.g., to -10 C). If the result is undesirable, the tissue can be rewarmed without permanent damage. Otherwise, the tissue can be permanently ablated by freezing it to a lower temperature (e.g., -73 C).

This therapy has revolutionized AV nodal reentrant tachycardia (AVNRT) and other AV nodal tachyarrhythmias. It has allowed people who were otherwise not a candidate for radiofrequency ablation to have a chance at having their problem cured. This technology was developed at The Montreal Heart institute in the late 1990s. The therapy was successfully adopted in Europe in 2001, and in the USA in 2004 following the "Frosty Trial".

In 2004, the technology was pioneered in the midwest United States at Miami Valley Hospital in Dayton, Ohio by Mark Krebs, MD, FACC, Matthew Hoskins, RN, BSN and Ken Peterman, RN, BSN. These electrophysiology experts were successful in curing the first 12 candidates in their facility. Many more have been successfully treated since and they continue their work in the field.

Cryoablation for AVNRT and other arrhythmias do have some drawbacks. A recent study concluded that procedure times are slightly higher on average for cryoablation than for traditional radio-frequency (heat-based) ablations. Also, higher rate of equipment failures were recorded using this technique. Finally, even though short term success rate is equivalent to RF treatments, cryoablation appears to have a significantly higher long term recurrence rate.

Site Testing

Cryotherapy is able to produce a temporary electrical block by cooling down the tissue believed to be conducting the arrhythmia. This allows the physician to make sure this is the right site before permanently disabling it. The ability to test a site in this way is referred to as site testing or cryomapping.

When ablating tissue near the AV node (a special conduction center that carries electrical impulses from the atria to the ventricles), there is a risk of producing heart block - that is, normal conduction from the atria cannot be transmitted to the ventricles. Freezing tissue near the AV node is less likely to provoke irreversible heart block than ablating it with heat.

Surgical Procedures

As in catheter-based procedures, techniques using heating or cooling temperatures may be used to treat arrhythmias during heart surgery. Techniques also exist where incisions are used in the open heart to interrupt abnormal electrical conduction (Maze procedure). Cryosurgery involves the use of freezing techniques for the treatment of arrhythmias during surgery.

A physician may recommend cryosurgery being used during the course of heart surgery as a secondary procedure to treat any arrhythmia that was present or that may appear during the primary openchest procedure. The most common heart operations in which cryosurgery may be used in this way are mitral valve repairs and coronary artery bypass grafting. During the procedure, a flexible cryoprobe is placed on or around the heart and delivers cold energy that disables tissue responsible for conducting the arrhythmia.

Cryoimmunotherapy

Cryoimmunotherapy is an oncological treatment for various cancers that combines cryoablation of tumor with immunotherapy treatment. In-vivo cryoablation of a tumor, alone, can induce an immunostimulatory, systemic anti-tumor response, resulting in a cancer vaccine - the abscopal effect. However, cryoablation alone may produce an insufficient immune response, depending on various factors, such as high freeze rate. Combining cryotherapy with immunotherapy enhances the immunostimulating response and has synergistic effects for cancer treatment.

History

The use of cold for pain relief and as an anti-inflammatory has been known since the time of Hippocrates (460-377 B.C). Since then there have been numerous accounts of ice used for pain relief including from the Ancient Egyptians and Avicenna of Persia (A.D.982–1070). Since 1899, Dr. Campbell White used refrigerants for treating a variety of conditions, including: lupus erythematosus, herpes zoster, chancroid, naevi, warts, varicose leg ulcers, carbuncles, carcinomas and epitheliomas. De Quervain successfully used of carbonic snow to treat bladder papillomas and bladder cancers in 1917. Dr Irving S Cooper, in 1913, progressed the field of cryotherapy by designing a liquid nitrogen probe capable of achieving temperatures of -196 °C, and utilizing it to treat of Parkinson's disease and previously inoperable cancer. Cooper's cryoprobe advanced the practice of cryotherapy, which led to growing interest and practice of cryotherapy. In 1964, Dr. Cahan successfully used his liquid nitrogen probe invention to treat uterine fibroids and cervical cancer. Cryotherapy continued to advance with Dr. Amoils developing a liquid nitrogen probe capable of achieving cooling expansion, in 1967.

With the technological cryoprobe advancements in the 1960s, came wider acceptance and practice of cryotherapy. Since the 1960s, liver, prostate, breast, bone, and other cancers were being treated with cryoablation in many parts of the world. Japanese physician Dr. Tanaka began treating metastatic breast cancer cryoablation in 1968. For the next three decades, Dr. Tanaka successfully treated small and localized as well as advanced and unresectable breast cancer with minimally invasive cryoablation. All of Dr. Tanaka's breast cancer cases were considered incurable: advanced, unresectable, and resistant to radiotherapy, chemotherapy, and endocrine therapy. At the same time, physicians, including Dr. Ablin and Dr. Gage, started utilizing cryoablation for the treatment of prostate and bone cancer.

The 1980s and 1990s saw dramatic advancement in apparatus and imaging techniques, with the introduction of CMS Cryoprobe, and Accuprobe. CT, MRI, ultrasound guided cryoprobes became available and improved the capabilites of cryoprobes in treatment. Excited by the latest advancements in cryotherapy, China embraced cryotherapy in the 1990s, to treat many oncological conditions. With the benefits well-established, the

FDA approved the treatment of prostate cancer with cryoablation in 1998. In 2003, American Radiologist, Dr. Littrup, began performing cryoablation on breast cancer. Dr. Littrup successfully performed cryoablation on breast cancer patients from stage I-IV, until the introduction of new Obamacare regulation in 2013, which ended the practice in the US.

References

- Peter J. Russel (2005). iGenetics: A Molecular Approach. San Francisco, California, United States of America: Pearson Education. ISBN 0-8053-4665-1.
- Rantala, Milgram, M., Arthur (1999). Cloning: For and Against. Chicago, Illinois: Carus Publishing Company. p. 1. ISBN 0-8126-9375-2.
- Watson JD (2007). Recombinant DNA: genes and genomes: a short course. San Francisco: W.H. Freeman. ISBN 0-7167-2866-4.
- Patten CL, Glick BR, Pasternak J (2009). Molecular Biotechnology: Principles and Applications of Recombinant DNA. Washington, D.C: ASM Press. ISBN 978-1-55581-498-4.
- Brown T (2006). Gene cloning and DNA analysis: an introduction. Cambridge, MA: Blackwell Pub. ISBN 978-1-4051-1121-8.
- Russell DW, Sambrook J (2001). Molecular cloning: a laboratory manual. Cold Spring Harbor, N.Y: Cold Spring Harbor Laboratory. ISBN 978-0-87969-576-7.
- David Price (2000). Legal and Ethical Aspects of Organ Transplantation – Google Books. Cambridge University Press. p. 316. ISBN 0-521-65164-6.
- David N. Weisstub; Guillermo Díaz Pintos, eds. (21 December 2007). Autonomy and Human Rights in Health Care: An International Perspective. Springer. p. 238. ISBN 978-1-4020-5840-0. Retrieved 21 May 2010.
- Nice.org Fertility: Assessment and Treatment for People with Fertility Problems. London: RCOG Press. 2004. ISBN 1-900364-97-2.
- Schulman, Joseph D. (2010) Robert G. Edwards – A Personal Viewpoint, CreateSpace Independent Publishing Platform, ISBN 1456320750.
- Professor Henry Louis Gates, Jr.; Professor Emmanuel Akyeampong; Mr. Steven J. Niven (2 February 2012). Dictionary of African Biography. OUP USA. pp. 25–. ISBN 978-0-19-538207-5.
- Zastrow, Mark (8 February 2016). "Inside the cloning factory that creates 500 new animals a day". New Scientist. Retrieved 23 February 2016.
- Shukman, David (14 January 2014) China cloning on an 'industrial scale' BBC News Science and Environment, Retrieved 27 February 2016
- Baer, Drake (8 September 2015). "This Korean lab has nearly perfected dog cloning, and that's just the start". Tech Insider, Innovation. Retrieved 27 February 2016.
- Ciralsky, Adam (January 31, 2016). "The Celebrity Surgeon Who Used Love, Money, and the Pope to Scam an NBC News Producer". Vanity Fair. Retrieved January 7, 2016.

Permissions

All chapters in this book are published with permission under the Creative Commons Attribution Share Alike License or equivalent. Every chapter published in this book has been scrutinized by our experts. Their significance has been extensively debated. The topics covered herein carry significant information for a comprehensive understanding. They may even be implemented as practical applications or may be referred to as a beginning point for further studies.

We would like to thank the editorial team for lending their expertise to make the book truly unique. They have played a crucial role in the development of this book. Without their invaluable contributions this book wouldn't have been possible. They have made vital efforts to compile up to date information on the varied aspects of this subject to make this book a valuable addition to the collection of many professionals and students.

This book was conceptualized with the vision of imparting up-to-date and integrated information in this field. To ensure the same, a matchless editorial board was set up. Every individual on the board went through rigorous rounds of assessment to prove their worth. After which they invested a large part of their time researching and compiling the most relevant data for our readers.

The editorial board has been involved in producing this book since its inception. They have spent rigorous hours researching and exploring the diverse topics which have resulted in the successful publishing of this book. They have passed on their knowledge of decades through this book. To expedite this challenging task, the publisher supported the team at every step. A small team of assistant editors was also appointed to further simplify the editing procedure and attain best results for the readers.

Apart from the editorial board, the designing team has also invested a significant amount of their time in understanding the subject and creating the most relevant covers. They scrutinized every image to scout for the most suitable representation of the subject and create an appropriate cover for the book.

The publishing team has been an ardent support to the editorial, designing and production team. Their endless efforts to recruit the best for this project, has resulted in the accomplishment of this book. They are a veteran in the field of academics and their pool of knowledge is as vast as their experience in printing. Their expertise and guidance has proved useful at every step. Their uncompromising quality standards have made this book an exceptional effort. Their encouragement from time to time has been an inspiration for everyone.

The publisher and the editorial board hope that this book will prove to be a valuable piece of knowledge for students, practitioners and scholars across the globe.

Index

A

Absolute Temperature Scales, 114

Absolute Zero, 110-117

Antifreeze, 2-3, 12, 55-56, 60-73, 78, 81, 106, 153

Antifreeze Protein, 2, 60, 64, 78

Applied Cryobiology, 3

B

Binding to Ice, 64

Biosecurity, 32-33

C

Chilblains, 133-134, 143-146

Cloning, 36, 163-181, 256

Cool Caps, 123

Cooling Catheters, 118, 120-121

Corrosion Inhibitors, 67-68, 71, 73

Cryoablation, 251-256

Cryobiology, 1-7, 10, 12, 14, 16, 18, 20, 22, 24, 26, 28, 30, 32, 34, 36, 38, 40, 42, 44, 46, 48, 50-52, 54, 56-58, 62, 64, 66, 68, 70, 72, 74, 76, 78, 80, 82, 84, 86, 88, 90, 92, 94, 96, 98, 100, 102, 104, 106, 108, 110, 112, 114, 116, 118, 120, 122, 124, 128, 130, 132, 134, 136, 138, 140, 142, 144, 146, 148, 150, 152, 154, 156, 158, 160, 162-256

Cryoconservation of Animal Genetic Resources, 9, 25, 38, 51, 56, 83

Cryogenic Deburring, 90

Cryogenic Deflashing, 87-90

Cryogenic Energy Storage, 98

Cryogenic Fluids, 83, 91-92

Cryogenic Fuel, 85, 95

Cryogenic Hardening, 84, 89

Cryogenic Machining, 90

Cryogenic Processing, 84, 91

Cryogenic Rolling, 90

Cryogenic Seal, 91, 94

Cryogenic Treatment, 89

Cryogenics, 1, 38, 82-87, 89, 91-93, 95, 97, 99, 101, 103, 105, 107, 109, 111, 113, 115, 117, 119, 121, 123, 125

Cryopreservation, 1, 3-26, 29-31, 33-39, 41, 43, 45, 47, 49, 51-53, 55, 57, 59-61, 63, 65, 67, 69, 71, 73, 75, 77, 79, 81-82, 225, 227, 229-230, 241-242, 244, 247

Cryoprotectant, 5, 10, 13, 17-18, 20, 31, 55-56, 58-60, 70, 77, 209

Cryostasis (clathrate Hydrates), 57

Cryosurgery, 1, 4, 6, 57, 65, 83, 163, 248-251, 254

Cryotank, 109-110

Crystal, 12, 34, 45-46, 50, 56-57, 60, 66, 70, 78-79, 88, 100-109, 111-112, 147, 153

Crystallization, 9, 12, 42, 48-50, 100, 104-107, 143

Crystallography, 100, 103, 108-109

E

Ectothermic Cooling, 152

Ectothermy, 151

Embryo Cryopreservation, 13, 23, 225, 227, 229-230, 241-242

Embryo Transfer, 21, 23-24, 36, 168, 218, 221, 225-227, 231, 240-248

Endothermy, 151, 153-154

Engine Combustion, 96

Ethylene Glycol, 23, 56-57, 60, 65, 68-71, 77, 248

Ex Situ Conservation, 26, 50-51, 53-55

F

Field Gene Banking, 51

Fish Afps, 61-62

Freeze Avoidance, 61, 75-76, 80

Freeze Tolerance, 9, 61, 65, 75-76, 78, 80

Freezing, 1-6, 8-24, 26, 28-29, 31, 33-34, 37-38, 42, 49, 55-57, 60-62, 64-68, 70-71, 74-80, 84, 86-87, 91, 102, 110, 116, 134, 137, 143, 145-147, 149, 153, 206-207, 209, 213, 248-254

Frostbite, 133-134, 143, 145-149

G

Glass Transition, 10, 12, 41-49, 55, 59

Glass Transition Temperature, 43-47, 49, 55

Glycerol, 5, 10, 12, 16-17, 56-57, 70, 77, 79, 81, 213, 248

Grid Energy Storage, 98

H

Heterothermy, 126, 130-131

Hibernaculum (zoology), 132

Hibernation, 1, 6, 76, 80, 126-131, 143, 158

Hybrid Organic Acid Technology, 72-73

Hypothermia, 4, 6, 59, 65, 118-121, 123-125, 130, 133-143, 145, 147, 149, 151, 153, 155, 157, 159, 161

I

Ice Nucleators, 2, 61, 77, 79

In Vitro Fertilisation, 12, 14, 23, 216-217, 222, 224, 227, 236-237

Indium Seals, 92-93

Insect Afps, 63

Insect Winter Ecology, 2, 75

Inter Situ, 52

Intracellular Freezing, 2, 10, 12, 80

L

Liquefied Natural Gas, 84-85, 95-96

Liquid Nitrogen, 3-5, 8-11, 14, 16-17, 25-26, 31, 34, 36-41, 51, 56, 58, 83-84, 86-87, 89-91, 96, 98-100, 147, 172, 209, 249-250, 255

M

Methanol, 68-69

Molecular Cloning, 163-164, 175-181, 256

N

Natural Cryopreservation, 9

Neuropreservation, 58-59

O

Oocyte Cryopreservation, 13-14, 18-22, 229-230

Oocytes, 4-5, 12-15, 19-23, 25-26, 29-30, 56, 219, 229-230

Organ Transplantation, 163, 183, 185, 189, 193-197, 203-204, 256

Organic Acid Technology, 72-73

Ovarian Tissue Cryopreservation, 13-14

P

Pathophysiology, 138, 147

Plant Afps, 62

Polymers, 42, 44, 47, 59, 107

Psychrophile, 73-74

R

Refreezing, 18, 148

S

Sea Ice Organisms Afps, 63

Seed Banking, 51

Semen Cryopreservation, 13, 17

Semen Extender, 209, 213

Silica, Sio2, 46

Slow Programmable Freezing, 11, 18, 23, 248

Social Egg Freezing, 22

Somatic Cells, 26, 30, 166, 168

Sperm Bank, 204-213

T

Targeted Temperature Management, 118, 120-122, 124-125

Thawing, 1, 4-5, 10-11, 16-18, 23, 25, 66, 206, 214, 236, 248

Thermal Hysteresis, 60-63, 65, 78

Thermoregulation, 131, 133-134, 136, 150-151, 154, 156-157

Third-party Reproduction, 204, 245

Tissue Culture, 51, 165

Trans Nasal Evaporative Cooling, 121

Trench Foot, 133-134, 143, 149-150

U

Uterine Preparation, 241

V

Vitrification, 5-6, 11-15, 18, 20-23, 29, 31, 41, 48-49, 56, 58, 229

W

Water Blankets, 122

CPSIA information can be obtained
at www.ICGtesting.com
Printed in the USA
LVOW05*1932311017
554464LV00004B/5/P

9 781635 490817